I0487493

The Jurassic

The Mammal Explosion

Georges P. Odier

The Jurassic
The Mammal Explosion

Copyright © 2006 Georges P. Odier.
All rights reserved.

Permission is granted to the Scientific Community for use of all the enclosed material
including text, graphics, and photographs. Full credit must be given.

All Graphics and Photographs by Georges P. Odier excepting those credited in the captions.

Printed by Trafford Publishing in Victoria, B.C. Canada

Note for Librarians: A cataloguing record for this book is available from Library and Archives
Canada at www.collectionscanada.ca/amicus/index-e.html
ISBN 1-4120-9013-X

Printed on paper with minimum 30% recycled fibre.
Trafford's print shop runs on "green energy" from solar, wind and other environmentally-friendly
power sources.

Offices in Canada, USA, Ireland and UK

Book sales for North America and international:
Trafford Publishing, 6E–2333 Government St.,
Victoria, BC V8T 4P4 CANADA
phone 250 383 6864 (toll-free 1 888 232 4444)
fax 250 383 6804; email to orders@trafford.com
Book sales in Europe:
Trafford Publishing (UK) Limited, 9 Park End Street, 2nd Floor
Oxford, UK OX1 1HH UNITED KINGDOM
phone 44 (0)1865 722 113 (local rate 0845 230 9601)
facsimile 44 (0)1865 722 868; info.uk@trafford.com
Order online at:
trafford.com/06-0769

10 9 8 7 6 5 4

In Memoriam

Francis 'Fran' Barnes' life was taken away by skin cancer in October 2003, a disease he had contacted during his 25 years of dedicated field research in Utah's Canyon Country. He was the first researcher to notice these strange 'tubes' in the Navajo Sandstone as far back as 1997, then raising questions about their origin. I teamed with him in 1999, then later proceeded to extend their scientific significance based upon Francis Barnes earlier speculation that they may be of mammalian origin. He passed away at the very time when science was beginning to acknowledge both these burrows and his impressive contributions to our better understanding of the Early Jurassic. I hereby dedicate this book to his memory.

Georges P. Odier

Table of Contents

Technical Information

Photographs

All photographs are by the author. Ruler in pictures is 6 in. long (15 cm). Coin is a 25 cent U.S. 'quarter'. Tracks are sometime wetted for better visual acquisition.

Graphics

All tracks, trackways, and burrow entrances are actual size. They are tracings of actual tracks, etc., made 'in situ' using translucent plastic sheet, a fairly new technique developed to study and compare track morphologies.

Location of sites

Actual locations are not given to protect them from vandalism. They are available however to anyone in the scientific community.

Note: My earlier book titled, The Jurassic – A new Beginning?, 2003, also published by Trafford Publishing, Victoria, BC, Canada, is at times referred to as my earlier book for simplicity sake. That book contains some photographs and graphics that were not duplicated in this document due to the cost involved.

Introduction

In my earlier 2003 book, "The Jurassic – A New Beginning?", also published by Trafford Publications, I then dealt with unknown discoveries affecting the conventional view of the Early Jurassic. This book is a sequel solely dealing with the survival and evolution of our Jurassic ancestors.

A discovery of major proportion was made in the late 1990's in the Moab region of Utah by the late Francis 'Fran' Barnes, later supported by my own, although at that time we didn't know for sure what we had uncovered. By the early 2000's, that discovery had expanded enough to raise some serious questions about its identity. Due to the general lack of interest from academic circles, it was put on the back-burner of field research until mid-2003, when the discovery of several mammal track sites brought it back to the front-burner. Its true identity finally surfaced in early 2004. What we had uncovered back in 1997 and 1999, was the then unknown habitat of our Jurassic ancestors.

Multitudes of burrows made by early mammals, thought to have been rare to extremely rare in the Jurassic and the Cretaceous. A discovery challenging the conventional view of these two periods, from the evolution of the mammal lineage to the dominance of the dinosaurs.

Although anchored in scientific data, this book's perspective is analytical, my field, from which a larger picture and implications can be drawn, albeit with some reserve since the documentation of these burrows has barely begun. These mammal burrows did not exist until now in the paleontological-geological record of the Jurassic-Cretaceous, and barely exist in the Late Cenozoic. Lacking precedent, search for evidence went beyond the constraints of paleontology and into inter-disciplinary studies. A subject that is also the objective of this book.

This book recognizes the difference between academic studies and projections derived from an assemblage of poorly documented discoveries. One can disagree with such projections until academic studies prove or support every step of the interpretative process. However, no one can disagree with, nor dismiss these discoveries now open to anyone. The goal of this book is not to embark upon arcane debates about the validity of every step but to paint a general picture sufficient enough to stimulate greater interest in our Jurassic ancestors.

Lastly, as an English-speaking person you may find the text a bit 'halting', not a polished product normally expected from such books. Please bear with it, I am French.

Georges Odier
Moab, Utah
August 15, 2006

Preface

Being a biologist, I must admit that I had only a passing interest in paleontology and left it to the 'experts'. I remember the controversy surrounding the K2 extinction of the dinosaurs and was happy to see the meteor impact theory finally accepted. What I found puzzling was the meteoric takeover by the mammals. It seemed too convenient, that such vast numbers with such diversity, could evolve from the small insignificant creatures, the mammals, as the paleontologists claimed. When George Odier introduced me to the burrows that are the subject of this book, what amazed me was, not that they were in the Jurassic, where they were not supposed to be, but the sheer number of them and the area they covered. My suspicion, that mammals existed in larger numbers and with greater diversity than the paleontologists maintained, seemed justified.

As a former science educator, it is important to me that the real story is told. I am sure we are all aware, that evolutionary theory is under attack from the religious right and any inconsistency in our account will be seized upon, with a view of discrediting the underlying principle on which current theory is based. It is important that research is continued in the true spirit of scientific inquiry, even if the evidence challenges long held theoretical systems. To create dogma is to play the same game as those that attack us. I was always taught that a theory was not a truly scientific one, unless there was the possibility that it could be refuted by appropriate evidence. Not only that, it was our duty to attempt to do so. Only recording evidence that supports the contemporary, conventional model, and ignoring that which may conflict is not true scientific inquiry. Sometimes, this may happen accidentally; Researchers simply do not see what is in front of them. This is understandable, but some things are not. Simply ignoring phenomena that have been brought to one's awareness, on the grounds that they don't fit the accepted view, is bad science and scientifically unethical.

The burrows, which this author, G.P.Odier, attributes to early Jurassic mammals, are a case in point. They have been clearly visible, for many years, but their origin was never investigated. This area of Utah is one of the most explored in the country, yet thousands walked past these phenomena, without considering them worthy of further study. After all, this area was the stomping ground of the great dinosaurs and these were not dinosaur bones, and so could not be of importance to our understanding of the Jurassic ecology, even though they littered the landscape and covered many square miles. It took two men, Francis Barnes and George P.Odier, to see, that they must have some significance and to further investigate what that could be.

Mammal remains are rare. Their small size makes it unlikely that many will be found. There are remains dating back to the Triassic and beyond, so what were they up to during the Jurassic? It appears that they were thriving in large numbers, not rare and retiring as previously thought. Most were probably living a fossorial existence; although with water and trees there would be other niches that they could exploit. Although the skeletal remains are rare, many tracks have been found, in association with the burrows. The question that is legitimate to ask is: if mammals were so prolific and had diversified successfully, why did they remain so small for so long, while the dinosaurs grew so large. Light came to this question only recently. The oxygen content of the atmosphere was calculated to be about half what it was to become, so the high metabolic rates of the small mammals did not allow a significant size increase, until oxygen began to approach present day levels. During the course of the investigation laid out in this book two very important discoveries came to light. The discovery, in China, of complete skeletal remains of two Jurassic mammals was announced. Even more significant was the discovery of another skeleton, about 50 miles or so from the burrow casts recorded in this book. The specialized hollow teeth of this mammal indicate an insectivore similar to an armadillo.

This book is not able to reveal such succinct evidence as skeletons, so it relies on what are called 'proxies', e.g. footprints and burrow casts. These provide scientific evidence that the animals were present, even though their physical remains have not been found. Other proxies used in this book are de-watering pipes that confirm the presence of water, and termite burrows that indicate the presence of vegetation. It is clear that we can no longer rely solely on the traditional methods of paleontology. Taking the attitude that only finding a fossil animal proves its presence on the scene is no longer viable, also that the numbers of fossils found are in the same relationship to those of the living animals. Genetic techniques are advancing rapidly and are producing results that conflict with the traditional morphological ones, particularly in taxonomy. In an article in National Geographic (April 2003: *Is the Truth in the Bones or the Genes?*), it is reported that many paleontologists angrily reject the DNA findings, arguing there is a problem with the molecular clocks geneticists use to date their claims. Geneticists counter, that paleontology just hasn't found the appropriate fossils. Who is right is something to which we can look forward.

The question this book seeks to illumine is not whether mammals were present in the early Jurassic, that is a fact, but in what numbers were they present, what was the dominant life style and what effect would they have had on the ecology of the areas they inhabited. If they were indeed present in the huge numbers this research indicates, then the current view of the ecological system within which their habitats are located may need significant revision. There is still much to learn about these fascinating warm-blooded animals to which we owe our own very existence, and this book is offered as a contribution toward this end.

Colin Egan, Moab UT, June 2006

Colin Egan is a biologist and educator from Newcastle-on-the-Tyne, England. Currently based in Moab, Utah, he is a member of the original investigative team and one of the founders of Moab Interdisciplinary Research Associates.

Chapter I

The Forgotten Mammals

A brief profile of our Jurassic ancestors.

Brief indeed. The December 9, 2003 issue of the Salt Lake Tribune, Utah's main newspaper, pretty well summarizes the state of the Jurassic mammals. Under a three parts scientific series exploring the dinosaurs' origins and expansion, replete with arrays of dazzling illustrations and pertinent information (1), our ancestors were dealt with in the following manner: *"Mammals were still rare during that period, and were small, nocturnal insect eaters "*.

With such an explanation one can wonder if they even existed. They did exist however and the reason for their obscurity is two-folds: One, these rat-like species or worse, 'mammal-like reptiles', are of little interest to the general public, perhaps even repugnant to some as our ancestors, and two, their fossils are so rare and fragmented that their existence, physical attributes, and radiation is the domain of handfuls of dedicated specialists (2).

To be sure they not only existed but are grouped today into distinct (or somewhat distinct) classifications. For instance in the late Jurassic of Utah around fifty different species of small rodent-like mammals are present in the fossil record (3): Among others, multituberculates, symmetrodondids, triconodondids, etc., common species also present elsewhere in most of the world - all associated via a common 'rat-like' appearance (4). But in the Early and Middle Jurassic these early mammals are less than clear and are (somewhat) visualized as follows:

The *Sinoconodonts* are the oldest group, either from the very early, or late Early Jurassic (?). The earliest of true mammals and the period when the first jaw hinge evolved.

The *Morganucodonts*, also from the Early Jurassic and close relative of the Sinoconodonts, that had a tooth replacement pattern suggesting lactation.

Beyond these two Jurassic groups we move into Cretaceous types. Among them the Early Cretaceous *Eomaia scansoria* (125 mya) the earliest known protoplacental (along with feet-claws adapted for tree-climbing). Toothed monotremes of unclear dating that had modified tribosphenic molars, allowing for a varied diet. And a certain *Jeholodens* (?) dated at around 125 mya that had mobile shoulder girdle increasing the range of motion (5).

All these determinations were mostly if not solely, based on jaws and teeth configuration, sometime backed by fragmentary skeletal remains. The up-to-date list of complete fossils, repeat complete, is very short indeed. In order of discovery they are: *Kayentatherium*, a fairly large tritylodont. As the name implies, it was first uncovered in the Early Jurassic Kayenta Formation in-around the Moab region of Utah (6). That fossil was incomplete however. Another, also incomplete, was uncovered in the Navajo Sandstone of northern New Mexico in 1991 (7). Almost identical in size these two fossils combined were able to offer a nearly complete skeleton of this tritylodont. Also uncovered in the Kayenta Formation are the fragmented remains of two other tritylodonts, *Oligokyphus and Dinnebiton*, along with scattered remains of mammals, *Dinnetherium* and haramyids (mice-like animals). *Eomaia scansoria*, a recent Chinese discovery, is the only complete fossil of the Cretaceous group - a small morganucodontid whose Latin name translates into 'dawn mother', so-named because it is the first and only protoplacental mammal ever found to date. Then comes *Hadrocodium*, also a recent Chinese discovery. A tiny (3 in. long) fossil superbly preserved right down to its most minute physical details. *Hadrocodium* is dated at 195 mya (very Early Jurassic) and is the first complete true mammal fossil ever uncovered in the Jurassic anywhere.

That was then. In 2005, two more mammals were added to the list. The first one, from China, *Repenomamus gigantiticus*, and its smaller cousin *Repenomamus robustus*, the second, *Fruitafossor windscheffeli*, from Colorado. Three complete fossils. The first one, *R. Giganticus-Robustus*, from the early Cretaceous, is still shaking paleontology, particularly its higher ranks who had for decades informed, if not imposed upon other sciences that such animals did not exist in the Mesozoic. These two large specimens were not only mammals, but carnivorous and feeding on small dinosaurs. The second, *Fruitafossor*, also a mammal, is a small termite-eater from the Upper Jurassic (Morrison Fm. of the southeastern US). These two discoveries, including Early Jurassic mammal burrows recently uncovered in Utah, are today forcing a debate as to what these early mammals really were, but more important, their demographic extension from the beginning of the Jurassic onward.

This debate had been closed for decades, based upon the belief that mammals only achieved their numbers and diversity after the demise of the dinosaurs at the 65 mya K-T Boundary. This belief, still in vogue in most paleontological circles, is anchored in two juxtaposed ideas, one scientific, the other a bias against mammals largely fueled by the huge public interest toward dinosaurs, and career-advancement to match it. This bias has several aspects such as, 'publication bias', 'preservation-Interpretation bias', 'sedimentation bias', 'evolutionary bias', etc. (8).

The scientific aspect has its roots in the Upper Triassic. During that period, known as the 'Era of the Mammals', a vast array of primitive mammals became the dominant species, even though toward its end primitive dinosaurs had made their debut. These primitive Triassic mammals are grouped under one general term, *Therapsids*. Among them however were more advanced species, but in much lesser numbers. These are generally referred to as 'mammals'. These mammals had begun their diversification by the late Triassic and by the end of the Late Jurassic eight mammalian lineages are known from the fossil record. While most, if not all paleontologists are in accord regarding Therapsids and mammals, the debate begins after the Triassic-Jurassic Boundary, i.e, the Early Jurassic. This is where the bias against mammals comes into focus.

A number of paleontologists, particularly the higher ranks (but not all), proposed if not imposed, that most of the mammaloids who crossed the Triassic-Jurassic Boundary, then settled in the Jurassic, and later in the Cretaceous, were mammal-like reptiles, a term implying that these mammaloids were still 'reptilian' in nature, and not part of the Mammal Class. During the Jurassic-Cretaceous period, *'mammal-like reptiles'*, along with *Synapsids and Therapsids*, are still commonly found in conventional literature even though they are misleading and scientifically inaccurate. In a field dominated by Dinosaurs, persons who describe our Jurassic ancestors as *mammal-like reptiles* are the sole living proof of their existence. The real but covert purpose to classify these Jurassic mammals as 'mammal-like reptiles' is 'to keep mammals in their place' in order for them to conveniently 'explode' after the demise of the dinosaurs at the 65 million years K-T Boundary. Under this scientifically primitive proposition, carnivorous dinosaurs were keeping the mammal population in check, including 'mammal-like reptiles', by feeding on them as prey species. Following the demise of the dinosaurs, these supposedly rare mammals were then free to 'explode' in exuberant numbers. A nice and easy to visualize theory but a false one. Here is the up-to-date and *scientific* classification of mammals prior to and during Mesozoic times:

The *Synapsids* are the oldest of the mammalian lineage. These pre-mammalian *Synapsids* dominated the land vertebrate fauna of the Permian and Early Triassic before losing ground to the dinosaurs and other archosaurs. When first uncovered in the late 1800s, they were baptized 'mammal-like reptiles'. A term reflecting the then extremely limited knowledge of prehistoric animals, but also the still prevailing rejection of Darwin's Theory of Evolution. Although many had characteristics in common with mammals, none of them were actually reptiles. They were a branch of earlier amphibians known as *Ichthyostega*, a sister group of the *Anthracosaurs & Temnospondyls*. One of the other branches that stemmed from these earlier amphibians were the *Archosaurs* – from which *reptiles* originated. Out of the six groups of the pre-mammalian *Synapsids* only one survived, the *Mammalia Group*; the others, Pelycosaurs, Dinocephalia, Dicynodonta, basal Cynodonts, and Gorgonopsia, became extinct.

The *Therapsids* – the members of the Mammalia Group - were advanced *Synapsids* and are the direct descendants of mammals. These *Therapsids* became the dominant land animals during the Middle Permian and consisted of three major classifications: The Dinocephalians, the herbivorous Anomodonts, and the mostly carnivorous Biarmosuchians. Following the Permian-Triassic boundary extinction *Therapsids* were reduced to only one or two families of a few species each surviving into the Triassic. Of these, the *Dicynodonts* (only represented by a single family of large herbivores), the *Kannemeyeridae*, and the medium-sized *Cynodonts* (in both carnivorous and herbivorous forms). Most died-out by the Late Triassic. However, a derived and more advanced group of *Cynodonts* survived. Known as *Eucynodonts*, three sub-groups of them managed to cross the Triassic-Jurassic Boundary and are classified as follows:

1. *Tritheledontids*, an extremely mammal-like group. This group is unknown beyond the Early Jurassic.
2. *Tritylodontids*, also an extremely mammal-like group, managed to survive until the Early Cretaceous.
3. *Morganucodonts*, and similar mammals like the *Sinoconodonts*, are or appear to be the main mammal branch that survived until the 65 million year K-T Mass Extinction.

The above information, *solely based on the fossil record*, seems to offer a fairly accurate view of the mammal lineage from the Permian to the end of the Cretaceous. As a note, the only known *Cynodont* that managed to survive up and into the early part of the Jurassic was *Oligokyphus*, an herbivorous species. In conventional literature only the *Morganucodonts* (and similar animals), are classified as *true mammals*, while the others, the *Tritheledontids* and *Tritylodontids*, are referred to as *mammal-like reptiles*. Enter cladistics.

Until recently paleontology was dominated by a scientifically primitive methodology *solely based on the fossil record*. Unlike similar sciences like paleoanthropology, conventional paleontology had not progressed beyond the one-dimensional but was able to impose upon other scientists and the general Public projections and assertions without support or evidence from inter-disciplinary studies. This era is fortunately coming to an end. In recent years, genetics, statistics, studies of 'proxies', among others, are beginning to challenge propositions solely based on the one-dimensional. The latest addition to enter the scientific fray are Cladistics.

Here is a synopsis of this relatively new science: Cladistics is a particular method of hypothesizing relationships among organisms. The basic idea behind cladistics is that members of a group share a common evolutionary history, and are 'closely related', more so to members of the same group than to other organisms. These groups are recognized by sharing unique features which were not present in distant ancestors. These shared derived characteristics are called *synapomorphies*. In other words, cladistics analysis forms the basis for most modern systems of *biological classification* which seek to group organisms by *evolutionary relationship*.

Without even adding statistical analysis, zoology, biology, genetics, etc., cladistics are good enough by themselves to eliminate any further ideas of 'mammal-like reptiles', 'Synapsids' and 'Therapsids' in the Jurassic and beyond. The three sub-orders of the *Eucynodonts* that crossed the Triassic-Jurassic boundary – Tritheledontids, Tritylodontids, and Morganucodonts – were **mammals**, nothing else.

Anyone unfamiliar with the above information can verify their veracity by a quick search on your computer. No need to buy or borrow expensive scientific books, nor spend hours at your local library. Under Synapsids, Therapsids, etc., and various sciences like cladistics, genetics, etc., numbers of scientific and informative web-sites are available at the tip of your fingers.

With 'mammal-like reptiles' out of the way, something that was already being eliminated, but should have been years ago, we now proceed to consider, who among these three sub-orders, was the main and only mammal lineage that reached and crossed the K-T Boundary, 65 million years ago. According to the 'mammal-like reptile' supporters, only the 'rare' and diminutive *Morganucodonts* reached that Boundary, to 'explode' immediately afterwards into the amazing diversity and multitude of the modern Mammal Class. To believe in such a miracle is to believe in Santa Claus. Yet it was, and still is, the Christmas present these conventionals have shoveled down the Public throat for the past 100 years.

In reality, the past and current *fossil record* is insufficient by itself to assert, as it had done in the past, that the *Tritheledontids* and *Tritylodontids* – an/or similar species – had vanished out of the Jurassic without a trace, and without any explanations. Another convenient miracle by itself. This miracle was recently debunked by the arrival on the scene of *Eomia scansoria*, the two *Repenomamus*, *Giganticus* and *Robustus*, and the latest *Fruitafossor* windscheffeli, not to mention the earlier and lesser known *Hadrocodium*. This new array of mammals took the conventional paleontologists by surprise, especially the two *Repenomamus* – carnivorous mammals eating small dinosaurs – indicating that early mammals were more diversified and advanced than previously thought, and by inference, that the dependence, solely on the fossil record to project varieties and numbers of mammals, is in great need of review.

The discovery of gigantic mammal burrows in the Early Jurassic of Utah, including some in its Upper Middle Jurassic, further challenges the conventional concept that mammals were 'rare, nocturnal, and insectivores'. In fact, they were present by the millions, as confirmed by these burrows. The question today is not their numbers but to what sub-order of the Mesozoic Mammal Class they belonged to. This we won't know until their skeletons are uncovered in these numerous burrows. Meanwhile, we can readily assume that, regardless of what they were, they did not miraculously vanish into thin air. They were and remained, the most numerous terrestrial vertebrates throughout the entire Mesozoic Period. These multitudinous mammal burrows have falsified the conventional and convenient view of the dinosaurs-mammals relationship. Carnivorous dinosaurs did **not** keep the mammals' variety and numbers in check by feeding upon them. And since they were incapable to do so until the end of Early Jurassic, their chances to affect the mammal population in the vastly more secure and hospitable parts of the Upper Jurassic and beyond, must have been minimal. If in fact there were fluctuations in the mammals' variety and numbers, they must have come from other causes.

This multitude of burrows are 'proxies' for the lack or paucity of the fossil record, and when added to cladistics, statistical analysis, genetics, biology, etc., they brush aside the preposterous notion that the entire Mammal Class – repeat *entire*, from mice, elephants, humans to whales – solely stemmed from a handful of 'Three Blind Mice', the very small, shrew-like *Morganucodonts* who managed to survive the nasty but impotent carnivorous dinosaurs up to the very end of the Cretaceous. As you say in the English language, it's 'Time to get Real'.

As for the track record, it's not any better, if anything worse than the fossil one. Until recently the entire Jurassic was dominated by one type of track: *Brasilichnium*. This type of track, coined in Brazil as the name implies, is similar to the older *Laoporus* type first documented in the Permian but still unknown in the following Triassic Period (9). Beyond these fairly common *Brasilichnium-Laoporus* tracks, only two others are attributed to mammal-like animals : *Navahopus*, a larger *Brasilichnium*-type from Arizona, and *Ameghinichus*, very small tracks from Argentina that clearly were made by a small mammal. To this we add a trackway discovered and documented by Prof. Martin Lockley in 2003 near Denver, Colorado, in the Cretaceous, but still undetermined at this date. Outside *Brasilichnium* tracks, no other tracks attributed to mammal-like animals have been documented in the Jurassic of Utah until 2002. I could be wrong however. Other types of mammal tracks may have been uncovered elsewhere in the US or abroad. Being aware of that possibility I have chosen not to give a formal name to my own discoveries until other specialists confirm they are indeed new types of tracks - this to avoid the problem of 'Provincial Taxonomy' that has plagued the track record in the past (10).

All the above combined is how our Jurassic ancestors are being currently perceived in paleontological circles, and how the Salt Lake Tribune could print in total sincerity that these early mammals were 'rare, small, nocturnal, and insect eaters'. But this document is not about sincerity but about *reality*. So let's take a hard look at the Salt Lake Tribune's descriptions:

Rare : True, if solely based on the *fossil* record. Paleontologists are scientists not visionaries, thus the need for hard or plausible evidence to determine and defend any scientific positions. From the 'one jaw - one vote ' point-of-view early mammals in the Early - Middle Jurassic seem indeed rare to extremely rare.

But the 'one jaw - one vote' concept, also known as 'show me the fossil', is more than questionable when dealing with difficult and complex situations such as the Early - Middle Jurassic where solid fauna and geological evidence is either lacking or uncertain. When it becomes an end by itself, it's no longer a true scientific proposition, but a somewhat arbitrary means to reach convenient or pre-supposed conclusions (11). The case for the 'rare mammals' was clearly based on the extremely narrow 'one jaw - one vote' concept not on sophisticated projections - projections that should have been made or at least attempted, since related but disruptive information were already available.

We begin in the Triassic. This Period is heralded in most scientific literature as the 'Era of the Mammals'. No point here to describe why, nor paint a picture of these various 'mammal-like reptiles' or mammal-like something. While current studies point to a decline in the mammalian domination toward the end of the Triassic - due the beginning of the dinosaurs' rise, particularly the Theropods - there is no evidence whatsoever that this multitude of mammaloids had vanished at the Triassic-Jurassic boundary either through predation, climate change, intermittent extinctions, etc., or via the presumed massive extinction at the Triassic-Jurassic boundary (12).

The sole evidence then available was the disappearance, at or near that boundary, of certain types of small or large mammal-like species - evidence being the then track and fossil record, but nevertheless open to serious challenges, as this record did not include any projections or speculations, that some or many of these Triassic mammal-like species had (or may have), already evolved into true mammals prior to the Triassic-Jurassic Boundary. The body fossil record only indicates that true or close-to-being mammals had begun their diversification in the late Triassic. Clearly, the then track and fossil record was not sufficient to embark upon solid determinations of the mammal fauna at hand, especially its numbers - only educated projections could have provided a more plausible view of that era. None were made or attempted, at least to my knowledge, leaving the 'one jaw - one vote' concept as the sole denominator for the mammals' presence and numbers.

The demand for projections, speculations, or other types of investigations, usually goes along with the scarcity of fossil evidence (13). This is particularly true when dealing with mammal species. Every paleontologist and researcher knows that the preservation of mammal remains, more so the smaller ones, is tantamount to a miracle, more so in the arid Early - Middle Jurassic formations of the southwestern US. A reality backed by the scarcity, or even lack, of large *dinosaur* fossils although these species are known to be present via the track record. Many a time teeth (sometime complete with jaws) are the only part of the body that stands a chance of preservation, since they are generally (but not always), impervious to soil acidity, scavenging, and other forms of disappearance (14).

But to uncover mammals' small to tiny teeth - or jaws if one's is lucky - tantamount to the proverbial 'looking for a needle in a haystack'. Early mammals obviously did not share the same social circles as predatory Theropods - or other large dinosaurs - so the chances to find their fossils in dinosaur quarries, pits, or killing grounds, where most dinosaur skeletons are found today, are near zero. We only guessed that most of these early mammals must have lived in some kind of colony-like burrows. However nobody really knew how these burrows may have looked like since none were actually reported until 2001 (15). So, without any physical knowledge of these would-be burrows, field searchers by-passed them as geological anomalies.

We then proceed to the Late Jurassic. In Utah for instance we have slightly more than fifty documented species, so far from eight mammal lineages, most of them only known from teeth-jaws or fragmented remnants. In other words, after a 50 million years 'gap' of scarce or no fossils we land into a relative over-abundance of them. Appropriately, this took place in the Morrison Formation (or similar formations), the end of the 'desolate deserts' and the beginning of the temperate and hospitable Late Jurassic of Utah. This is where a literal explosion of life gave us, via a now prolific fossil record, an almost over-abundance of 'Jurassic Park' dinosaurs. This is also where early mammals or 'mammal-like reptiles', depending to whom you talk, made their timid entry onto the scene. But overwhelmed by the 'Hollywood dinosaurs', they were to remain in the background until the K-T Extinction. There, the demise of entertaining dinosaurs brought them back to the front burner.

The Late Jurassic is also where our ancestors acquired via the fossil record, a somewhat repugnant 'rat-like' appearance, further relegating them to obscurity. Their evolution from Triassic 'mammal-like reptiles' to Jurassic-Cretaceous 'rat-like' rodents certainly did not help their cause.

However 'rat-like' or cute they may have been, their Late Jurassic variety should have raised at least some concern about their presumed scarcity. By that time all these species - early monotremes, protoplacental, or whatever they may have been, is not important – what is important is that they shared a common mammal characteristic: *prolific litters*. Thus, this method of survival specific to mammals, especially prey species, should have been taken into account (as it was in the Triassic earlier on).

This lack of correlation, projection, or speculation bring us back to the original problems that have plagued early mammals from the Early Jurassic onward: One, over-dependence on a conventional and nearly sacrosanct geological record: presumed desolate deserts, inhospitable to any kind of life, water and vegetation; and second, over-dependence on a fossil and track record suggesting both a scarcity of mammals and a period dominated by predatory Theropods. A scenario commonly found in paleontology literature implying that Theropods were mainly or solely feeding on small mammals (when not feeding 'on each other'). A questionable proposition that conveniently kept mammal-like populations down to 'rare levels' reflected by the scarcity of fossils and tracks - keeping in mind that these mammal populations in the Early - Middle Jurassic were already deemed 'marginal' due to the also presumed severity of these desolate deserts.

In essence it was assumed that the conventional *geological* record had priority over the fossil and track record. This assumed priority, still in vogue today in many quarters, had tied the hands of too many paleontologists (and ichnologists), forcing a somewhat arbitrary 'adaptation' of the fauna at hand - especially mammals - to a rigid and many times dogmatic geologic record. Today, more so in the future, it's the other way around: the geological record must now adapt to or take into consideration the ever increasing track and fossil record to be considered scientifically solid. A change of venue unwelcome in some circles, as it upsets the conventional scientific supremacy of one discipline over another. I will discuss this matter in more depth in a later chapter.

To all the above we add the current and still minimal academic interest in early mammals, as discussed in the 'Tracking' vs. Field Research' side-bar. A factor reflected in field researches. Scarcity of mammal tracks and fossils is mainly due to two inter-locking causes: lack of field experience compounded by lack of interest. No wonder early mammals were so 'rare' and barely on the radar screen.

Small : No contest. Using the world-wide and up-to-date fossil record, the largest ever found in the Jurassic anywhere is *Kayentatherium* while the smallest is the recently discovered Chinese *Hadrocodium*. *Kayentatherium's* length is slightly over 4 feet (1.4 meters) with a shoulder height of 14 inches (35 cm). *Hadrocodium* on the opposite is approximately 3 inches long (8 cm) tail included. Further, the Late Jurassic-Cretaceous record doesn't show any species larger than say, a small contemporary dog. While this record is less than satisfactory due to many factors already discussed it appears to be in the ball park.

The up-to-date track record (including my recent discoveries) largely supports the fossil one. Large but rare tracks of *Brasilichnium* configuration, tentatively attributed to some kind of 'mammal-like reptiles' - or large early mammals? - are present up to the end of the Early Jurassic of southeastern Utah, but not beyond. Diminutive tracks, including a set that may be the smallest ever found, were uncovered at the very beginning of the Jurassic (Lower Wingate) but their determination is still pending at this date. In the Kayenta Formation, more so in the Navajo Sandstone, a fairly large number of *Brasilichnium* tracks, including two new types still under investigation, are small to very small in size suggesting equivalent trackmakers. While this assortment of tracks supports a variety of small early mammals, a very recent discovery in the Moab-Entrada Member of the Entrada Sandstone (the final deposit of the Middle Jurassic in Utah) confirms that larger species (perhaps as large as a contemporary small coyote) were also present at that geologic time, thus may have been present in the Late Jurassic and earlier. The discovery of *Repenomamus Giganticus* in the Early Cretaceous supports that projection. More about these tracks in the next chapter.

But small can also be interpreted as insignificant. When followed by 'nocturnal' as it does in the Salt Lake Tribune's description we are not dealing here with actual measurements but a suggestion of *unimportance*. In other words a diminutive and obscure lot of 'rat-like' prey species - basically uninteresting 'fodder' for the vastly more entertaining Theropods now center-stage at multitude of 'Dinosaur Museums'. As suggested by the Royal British Columbia Museum's recent exhibit of Chinese dinosaurs we have a long way to go before ever seeing a 'Jurassic Mammal Museum' anywhere.

Nocturnal : The finishing touch that was needed to make these already rare mammals disappear from view. The best I can offer on the nocturnal 'thing' are: 1. Day temperatures *anywhere* in the gigantic Pangaea Supercontinent were so intense that our ancestors could only forage at night, and 2. They were so 'rare' that in order to escape these multitudes of hungry Theropods they could only come out of their holes during moonless nights. Indeed some imagination is required. More so since in fact they were present by the millions. What's yours?

Insect eaters : Here we are back on solid ground : conventional geology had declared the Early and Middle Jurassic of Utah a series of 'desolate, inhospitable deserts' devoid of water, vegetation, and other amenities, thus per force these small, rare, and nocturnal mammals had to eat and drink something else beside sand. Insects, preferably large, abundant, and juicy, were clearly the answer. No one has yet to explain what these multitudes of insects could have possibly fed upon themselves: sand, each other, or vegetation? The later being the only but annoying answer you will find plenty of insects in these 'forbidding deserts' but no plausible explanations for their existence as this would upset the conventional geological applecart. In these make-believe deserts dinosaurs, mammals, and insects do NOT equal water or vegetation, which is the same, but imaginative or arbitrary formulas designed to perpetuate conventional ideas. But behind such questionable propositions, hides something vastly more serious: *Who were these early mammals ?*

To show how imprecise our view of these animals still is I will quote the following from Frank de Courten's "Dinosaurs of Utah", 1998, currently the most comprehensive study of the Mesozoic fauna in this region:

" The advanced herbivorous mammal-like reptiles known as tritylodonts were especially abundant and diverse during Kayenta time (middle Early Jurassic in southeastern Utah). *Kayentatherium* is the most common of this group and may be close to the ancestor of late Mesozoic mammals. Some early members of those advanced mammal groups are known from the Kayenta Formation. *Morganucodon* and *Dinnetherium* (16), for example, both have been uncovered from Arizona localities and are so far advanced on the evolutionary path to becoming mammals that most paleontologists place them into that class without question ".

This short paragraph pretty well summarizes how these early mammals are still perceived in most (but not all) paleontological circles: Abundant and herbivorous 'mammal-like reptiles' (to date only backed by two fairly complete fossils and a handful of tracks), and advanced mammal groups only known from teeth and fragmentary skull remains. Hereby the origin of the 'rare, small, nocturnal, and insect eaters' proposition. Confusing to say the least, with *herbivorous* and 'abundant mammal-like reptiles', who, a short Boundary away, were herbivorous, omnivorous, and carnivorous, and 'rare' early mammals arbitrarily described as 'insectivores' without supportive inter-disciplinary evidence of such restrictive and evolutionary-limited diet.

The 'insect eaters' proposition is of the most interest since by definition insectivorous means *omnivorous* - partial or occasional flesh eaters that again translates into partially *carnivorous*. To what degree these early mammals were herbivorous as opposed to carnivorous is a key to their true profile and methods of survival.

Herbivorous, insectivorous, or omnivorous classifications were mainly determined from teeth studies (17), since most of these early mammals are only known from their teeth. However these classifications are arbitrary, unless integrated into a much wider view of these animals - a great deal outside the confines of paleontology. In zoological terms most of these small early mammals were *rodents* with teeth configuration similar or fairly close to contemporary species. And in rodents, teeth configurations can apply to anything from solely herbivorous to partially carnivorous (omnivorous) depending on the habitat, place on the food chain, and methods of survival.

Thus to classify early mammals as 'insectivores' or 'herbivorous' solely based on teeth configuration is factually inaccurate, and by extension arbitrarily imposes habitats, methods of survival, etc., without providing any proof of their existence. In contemporary mammals only moles (Order Insectivora) and termite/ant eaters like armadillos (Order Edentata), manis (Order Pholidota) and aardvarks (Order Tubulidentata) have reduced teeth connected to an entirely insectivorous diet. In Mesozoic times, the only known early mammal (including 'mammal-like reptiles') with such teeth and diet characteristics is the brand new fossil *Fruitafossor*, uncovered in 2005 in the Late Jurassic of Colorado. In the following chapters I will discuss new discoveries and how they affect the conventional perception of these early mammals. In the meantime here is a very personal and fairly accurate view of our ancestors:

Upon a strong suggestion by Canadian colleagues, I found myself on May 12, 2003 at the Royal British Columbia Museum, Victoria, BC, Canada, looking at the first-ever Chinese exhibit of dinosaur fossils, heralded as the 'Dragon Bones' (18). However I did not drive 3,000 miles to stare at fairly common dinosaur fossils, albeit well-preserved or Chinese they may be, but to have a first-hand and close view of mammal fossils in particular the very Early Jurassic *Hadrocodium*, the early Cretaceous *Eomaia scansoria*, and hopefully other mammal fossils still unknown in western scientific circles.

What I found was worth my long journey but not what I was looking for. Among a multitude of amazingly well-preserved skeletons, I 'discovered' two still largely unknown dinosaurs (in the US and Canada) of huge implications: *Lufengosaurus*, a very large prosauropod from the Early Jurassic Period (205 - 200 mya), and *Bellusaurus*, a small but true Sauropod dated at 170 mya, confirming that such species were indeed present in the Early-Middle Jurassic (19). Then in a large adjacent room were *all* the current fossils of 'dino-birds' and early birds discovered in China. A truly amazing display. All and all a staggering amount of fossils and information worth a 3,000 miles trip. But where were the mammals?

They were there of course and in fact properly represented in accordance with their current importance: Between a huge Sauropod called *Mamenchisaurus* - so huge that the Museum had to remove the false ceiling to incorporate its long neck - and a fierce-looking Theropod by the name of *Monolophosaurus* was a very small stand with a huge magnifying glass suggesting something extremely small and of possible scientific interest to somebody. This is where I found the sole representative of our Jurassic ancestors: *a tiny jaw* (20)

But you won't need a magnifying glass to find them. By the end of this document they hopefully will come into better focus.

Side-bar

Mammal-like Reptiles? What does that mean?

This title, written by Tamar Simon for the Discovery Channel web-site, symbolizes the ambiguity of the term. Who coined it is unknown, but in the English language it first appeared in a naturalist paper dated 1895 (London, England), describing a, new or not, fossil named *Dimetrodon* (then spelled *Demetriodon*). Although potentially incorrect that's the best my research could do. Correct or not, from that period onward the term 'mammal-like reptiles' has remained to describe the origin of mammals in most scientific literature, but not all. In Europe for instance, particularly in secular countries like France where religion and science are severely kept apart, 'mammal-like reptiles' are either discouraged or banished from scientific literature, usually replaced by 'Mammaloids', a proper scientific name for the earlier forms of the mammal lineage.

As paleontology evolved with increasing numbers of fossils, the mammal lineage began to surface in the Permian Period, among a mind-boggling array of reptile-looking terrestrial animals. In its early days paleontology was incapable to differentiate reptiles from amphibians, or for that matter, the various branches that had evolved or were evolving from the original amphibians. Thus, 'mammal-like reptiles' became a 'generic' term to describe the earliest known branch of the mammal lineage: the *Synapsids*. From there on the term stuck. The more evolved *Therapsids*, then the *Cynodonts*, then again the *Eucynodonts* of the Late Triassic and Jurassic, are still largely described as 'mammal-like reptiles'. A term not considered a formal one by most experts, for technically speaking, the *Synapsids* were *not reptilian* but a branch of the Amphibians – from which the reptilian branch also stems. The proper name for this earliest branch of the mammal lineage should be 'Mammal-like Synapsids' or 'Mammal-like *Amphibians*'. So why use the term 'mammal-like reptiles' when they are not? Especially in the Late Triassic and Jurassic when *Cynodonts*, more so *Eucynodonts*, had already evolved into true mammals.

The answer is two fold: In 1895, paleontology, then known as Natural Science, was still struggling under the social and religious cloud rejecting Darwin's Theory of Evolution. Reptiles and dinosaurs fine, but no ancient animals suggesting any connections with human beings. Science however had already progressed beyond Creationism but not enough to stand on its own; thus the term 'mammal-like reptiles'. Reptilian enough to ward off social and religious pressure while introducing a timid 'mammal' into the higher circles of science.

The second answer is more complicated since it no longer deals with Creationist nonsense. It deals with the one-dimensional notion that true mammals only existed in minimal numbers until the demise of the dinosaurs. A notion solely based on the body fossil record. And to insure the supremacy of the fossil record over anything else, the term 'mammal-like reptiles' was applied to any kind of early mammals from the Late Triassic onward, except for a convenient sub-group of field-mice named *Morganucodonts*. With the Tritheledontids and Tritylodontids arbitrarily baptized 'mammal-like reptiles', read reptilians, the mice-like *Morganucodonts* were then safe to assume the sole ancestry of the Mammal Class following the K-T Mass Extinction.

But why call these Jurassic-Cretaceous early mammals 'mammal-like reptiles' since they were not? Paleontologists, like anyone, can read the world-wide scientific literature dealing with the mammal ancestry, more so since they are supposed to be the leaders in that field. They can read it all right, but most are unwilling to concede the supremacy of paleontology to related sciences. In other words, their 'mammal-like reptiles' have little to do with scientific reality only there to impose upon the Public and other sciences the narrowness of the body fossil methodology. Hereby, the *Morganucodonts* field-mice were our sole ancestors, simply because no one had bothered to look for the remains of the other groups at, or near the K-T Boundary.

The 'body count', as it is called in other scientific circles, is somewhat similar to our 'victory' in Viet-Nam. To count them you first have to find them. And in the US at least no one is specifically looking for them, nor their tracks, burrows, and other signs of their presence. To date, I am one of the few who does. No wonder mammals were 'rare, nocturnal, and insect eaters', or 'reptilian' if they don't fit in the conventional program. These days, China is the only country with a 'mammal program', and no surprise, finding them. Most discoveries of mammals and birds come from China, and their recent finds have shaken the entire US paleontological community.

'Mammal-like reptiles' were an oxymoron whose time has long passed. Nowadays, for anyone to use that term is to identify yourself with a primitive methodology, not the sophisticated and demanding aspects of modern science.

Chapter index

1. Tribune's sources: The Macmillan Illustrated Dictionary of Dinosaurs and Prehistoric Animals, New York; DK Dinosaur Encyclopedia, Dorling Kindersley, New York. On December 10, 2003 via an email I informed the Salt Lake Tribune's scientific editor that such an explanation was in serious need of revision based on recent mammal discoveries in the very State of Utah. He has yet to answer. As a rule, scientific information published in newspapers and magazines must be supported by scientific institutions or qualified individuals. In this case, the Salt Lake Tribune's explanation was not challenged by either BYU or the University of Utah, nor by the State Paleontologist or other professionals within that State, suggesting, if not confirming that it still represents the current scientific view of the Jurassic mammals. In Utah, at least.

2. Dedicated specialists, because still today there is very little public and financial support for the studies of early mammals. Their dedication must be praised because career advancements depend largely on spectacular or 'mediatic' discoveries not on obscure studies of little interest to anyone except small scientific circles.

3. Frank de Courten. 'Dinosaurs of Utah', 1998.

4. Baker, Carpenter, Clemens, among others.

5. Rick Gore, 'The Rise of Mammals', National Geographic, April 2003 issue.

6. Frank de Courten, 'Dinosaurs of Utah', 1988

7. It was discovered by Dale A. Winkler, Museum of Paleontology, Dallas, Texas, and documented in his following paper, 'Life in a sand sea: Biota from Jurassic interdunes', Sept. 1991.

8. These various biases were first reported, and explained, by M. R. Voorhies in his 'Vertebrate Burrows' study published under 'The Study of Trace Fossils' by Springer-Verlag in 1975.

9. Martin Lockley & Adrian Hunt, 'Dinosaurs tracks', 1995.

10. A serious problem pointed out by Lockley & Hunt in 'Dinosaurs Tracks', 1998

11. J. Lawrence Powell, 'Night comes to the Cretaceous', 1999

12. 'Assessing the record and causes of Late Triassic extinctions', article in press, 2003, by L. Tanner, S. Lucas, and M. Chapman. Then, 'Definition of the Triassic-Jurassic Boundary', 2005, Albertiana 32, by S. Lucas, J. Guex, L. Tanner, D. Taylor, W. Kuerschner, V. Atudorei, and A. Bartolini. In depth studies challenging the presumed TJB massive extinction, strongly supported by new fossil and track discoveries, including my own.

13. Stephen (Steve) Hasiotis (Assoc. Professor, U. of Kansas) was the first to note the importance of fossil burrows as 'proxies' for organisms unknown or little known in the fossil record, or only known via their tracks ('Complex ichnofossils of solitary and social soil organisms', published

on that date by Palaeo-Elsevier in 2002). However, the use of 'proxies' in paleontological determinations has yet to gain the recognition they deserve. Hasiotis was also the first to include burrows (any kind of burrows) under Ichnology, something many ichnologists, including some of the finest still refuse to do, sticking solely to tracks determinations.

14. Frank de Courten, 'Dinosaurs of Utah', 1988

15. As the one who, along with Francis 'Fran' Barnes of Moab, Utah, put mammal burrows 'on the map' here is a personal view of the matter: Although burrows were sometime (but rarely) mentioned in paleontology literature, they were mostly guesses. To my knowledge, Francis 'Fran' Barnes was the first to report, in 2001, a number of strange and unknown burrows located in the Navajo Sandstone. Prior to this date, in 1999, I reported similar finds, also in the Navajo, to the BLM regional paleontologist but suggested that they may be fossilized roots. Similar (?) burrows were reported by Anthony Martin (University of Georgia) in 2002 (?) in the Late Cretaceous of Montana. This information came from Evelyn Boswell (7/23/02) indicating that these burrows were still undocumented at that time. My requests for further information were not answered to date, so I am not sure of their status. In late November 2005, and unknown to me and my colleagues until then, Steve Hasiotis (University of Kansas) informed me of mammal burrows that he and his associates had documented in 2004 in the late Triassic and late Jurassic. This newly found information is discussed in details elsewhere in this book.

16. *Morganucodon* and *Dinnetherium* are nearly identical, and related to the *Sinoconodonts*. Shrew-sized early mammals only known from teeth and skull fragments. There are numbers of names for them in various part of the world where they have also been uncovered. *Dinnetherium* was uncovered in Arizona in 1983, *Morganucodon* much earlier but date unknown. There exist only 4 skulls of *Morganucodonts* throughout the world (1994). Most of the *Morganucodonts* teeth and skull fragments come from the Glamorgan Quarry in England, thus the name.

17. Fossil teeth studies are highly specialized and demanding. I will note here three paleontologists whose contributions in that field are of particular scientific value: Richard 'Rich' Cifelli, Oklahoma Museum of Natural History, Andre Wyss, University of California at Santa Barbara, and Zhe - Xi Luo, Carnegie Museum of Natural History in Pittsburgh.

18. That unique exhibit was put together by the Institute of Vertebrate Paleontology and Paleoanthropology of Beijing, People's Republic of China, and Dr. Richard Hebda, Curator of the RBCM, along with the help of provincial authorities. This Exhibit was then scheduled to remain in Canada but may have crossed into the US at a later date (?).

19. In my earlier book I dealt with this possibility following the discovery in 2002, in the Lower Wingate Formation, of tracks that can be attributed to early but true sauropods. Until the arrival of *Bellusaurus* and the above Wingate tracks, most paleontologists had rejected the possibility of true sauropods and ornithopods in the Early and Middle Jurassic, especially in southeastern Utah. Although the poor fossil and track record supported such rejections, projections of their presence, even in minimal numbers, should have been attempted. These two types of herbivorous dinosaurs had to be present in some form or another during these earlier periods since their fossils are very common at the onset of the Late Jurassic (Morrison Fm.). If not, then migrations from adjacent

territories should be explained - not as arbitrary cover-all 'migrations' but offering supporting evidence for such theories.

20. Actually the right segment of a jaw, 1 in. or 2.5 cm long, extremely well-preserved with all teeth present, attributed to some kind of *morganucodontids* and dated as 'Upper Early Jurassic'.

Chapter II

The mammal track record in the Moab region of Utah

Geology overview

The Moab area is located in southeastern Utah generally described as the Canyonlands or 'Canyon Country'. Its geologic stratigraphy ranges from the Upper Pennsylvanian to the Dakota Sandstone Formation of the middle-upper Cretaceous. However most of the Pennsylvanian, a major part of the Permian, including many horizons of the Triassic, are either difficult to access or their exposed areas badly damaged by erosion. The Navajo Sandstone (upper Early Jurassic) is the most exposed and the easiest to access. Vice-versa the Wingate Sandstone (lower Early Jurassic) is the least accessible of the Jurassic Period, mostly present via vertical cliffs. The Middle Jurassic is represented by the spectacular Entrada Sandstone, also known as 'Red Rock Country'. This period is mainly made of straight-walled 'mesas', but with scattered exposed areas, most badly damaged by erosion. However, the upper part of that deposit is partially made of a local 'tongue' known as the Moab-Entrada. This tongue is generally less eroded than the underlying Entrada Sandstone and contains fairly large exposed areas relatively easy to access. The Salt Wash and Brushy Members of the Morrison Formation (Late Jurassic) are generally scattered and in poor state of preservation. At the upper end of the geologic stratigraphy, only badly damaged remnants of the Cedar Mountain Formation (Early Cretaceous) and the upper Dakota Formation are present in this area.

Tracks overview

This region, as per all regions of the Jurassic, is dominated by one type of track named *Brasilichnium*, sometime referred to as *Laoporus*. For the uninitiated, this type of track can be summarized as follows: A track-type nearly identical to the tracks of contemporary species such as small dogs or felines. It was first uncovered in the Permian Period, is still unknown in the Triassic, then re-appears at the beginning of the Jurassic. By the end of the Early Jurassic (Navajo Sandstone) they are present in relative numbers, but not beyond. In the middle and late Jurassic they are either extremely rare or missing. Please keep in mind that in southeastern Utah the first track of anything ever reported was by Edwin McKnight in 1940 (a Theropod track).

This track-type was first documented in Europe in 1850 (in a Permian deposit), and then named *Chelichnus* (meaning ' tortoise tracks'). It was re-named *Laoporus* in 1918 by R.S Lull, an American paleontologist who first uncovered and documented these tracks in North America (in the Grand Canyon area of Arizona). In recent years they were again uncovered, this time in the Early Jurassic of South America, Brazil, by the famed priest-scientist Guiseppe Leonardi, who in 1981 gave them a new name: *Brasilichnium*. This is the name under which they are mostly known in America today, although a debate is still going on regarding their lineage or association with the original *Laoporus*.

This debate however is far from being academic. This type of track is the only one known today linking Jurassic mammals to Permian 'mammal-like reptiles' (*Therapsids*), and by inference suggesting an unbroken lineage all the way from that distant period to contemporary times.

Although many scientists both in America and abroad have tried (and are still trying) to unfold its mysteries. I will note two of them whose work on the subject are of particular interest: Martin Lockley, a professor of ichnology at the University of Colorado (1), and Spencer Lucas, the curator of paleontology and geology at the New Mexico Museum of Natural History (2).

Before we move on to the track record here are six pieces of information of importance to its comprehension:

1. In the entire Triassic the only tracks currently associated with 'mammal-like reptiles' are *Therapsipus* and *Dicynodontipus* (3). The first one, from Arizona, is of importance. It was recently uncovered in the Moenkopi Formation, the lower half of the Triassic in this region. Attributed to large, wide-bodied *Therapsids* – thus the name *Therapsipus* - they are the first ever found in that formation. In configuration these large tracks clearly resemble the much smaller *Brasilichnium* type and the similar but larger *Navahopus* type (see # 3 for details of that track). The *Therapsipus* trackways for the first time infer a morphological continuity in the types of tracks attributed to *Therapsids* from the Permian to the Jurassic. The second one, from Europe, is attributed to a very small 'mammal-like reptile' (*Therapsid*) whose tracks bear a resemblance to the *Brasilichnium* type.

2. In the very Early Jurassic of the southwestern US the first-ever tracks associated with 'mammal-like reptiles' – *Therapsids* or early mammals - were documented by Prof. Lockley in 1994. This *Brasilichnium-Laoporus* trackway was discovered in the Moenave Formation, a regional and wetter deposit confined to the southern part of Utah corresponding geologically with the upper horizons of the Wingate Sandstone.

3. A type of track known as *Navahopus* was uncovered in northern Arizona in recent times. This track from the upper Early Jurassic (Navajo Sandstone) suggests a trackmaker the size of a large dog. However, a few more have been found since, and due to its high resemblance to the *Brasilichnium* type, except larger, its classification as a separate track is contested by many ichnologists, including myself.

4. Another type of track totally different than the *Brasilichnium-Laoporus,* and clearly associated with a mouse-like mammal, was uncovered recently in the middle Jurassic of Argentina, and named *Ameghinichnus* (although another identical to this one was uncovered earlier and named *Eopentapodiscus*). This type of track is nearly impossible to separate from the tracks of contemporary gerbils. None of these have been uncovered to date in North America.

5. In 2004, Martin Lockley, Spencer Lucas, Adrian Hunt and Robert Gaston documented and reported two sets of tracks (trackways), in the Lower Wingate Sandstone (very Early Jurassic) near Gateway, Colorado, a tiny settlement 35 miles east of Moab, Utah (as the crow flies). One set is attributed to a *Synapsid* (i.e. an early mammal. *Synapsids* became extinct in the Middle Triassic, *Therapsids* in the upper part of the Triassic. The only segment of the *Therapsids* that survived the Triassic-Jurassic boundary were the three groups of the derived *Eucynodonts*. These three groups were *mammals*, period. *Oligokyphus*, is the only known Therapsid *Cynodont* that managed to cross the Triassic-Jurassic boundary but became extinct shortly afterward). The other is attributed to either a mammal or a 'mammal-like' trackmaker, both made by a very small animal. In the

manuscript submitted to the journal *Ichnos* in 2004, the tracks themselves are described as follows: "Based on comparisons with *Laoporus* and *Brasilichnium*, it appears that the Wingate tracks are more similar to *Laoporus* in having less size differential (less heteropody) between manus and pes". For the ones interested in this discovery here is the title of the manuscript: "Ichnofaunas from Triassic-Jurassic Boundary Sequences of the Gateway area, Western Colorado: Implications for Faunal Composition and Correlations with Other Areas".

6. Elsewhere in the Canyonlands of Utah and Arizona, *Brasilichnium* tracks or trackways were documented by Martin Lockley and his team around the shore of Lake Powell, now accessible due to low water levels (1998, 1999, 2003-2005). Some are larger tracks essentially indistinguishable from the patterns seen in both *Brasilichnium* and *Navahopus* types (in the Navajo Sandstone). John Foster, Curator of Paleontology at the Museum of Western Colorado, in Grand Junction, CO, also reported in 2005 some *Brasilichnium* tracks in the Grand Staircase National Monument (also in the Navajo Sandstone).

Now to the mammal track record in-around the Moab, Utah, region. Exact location of sites is not given to protect these valuable resources from vandalism and commercial ventures still rampant in this region of Utah.

Prior to early 2000

No site name but known as the 'Two Slabs' discovery. 1968
Upper Navajo Sandstone - upper Early Jurassic, approx. 186 mya.
Discoverer: Lin Ottinger, an early 'tracker' from Moab

Actually one very large slab that was moved from its original site in the Sand Flats area, then broken in two, one part eventually going to the Dinosaur Museum of Western Colorado (Grand Junction, Colorado) the other to the Moab BLM office where it is still displayed outside its entrance (4). This discovery was unknown to scientists until 1972. From there on it was subjected to a series of interpretations notably by Utah paleontologists Jim Jensen and Lee Stokes then later (1989 to1993) by Prof. Lockley, an ichnologist. The later interpreted all tracks as *Brasilichnium* except for a trackway made by a small Theropod and a handful still tentatively interpreted as *Batrachopus*, of crocodilian origin (5). The crocodilian interpretation was later supported by Guiseppe Leonardi, K. Padian and P. Olsen, all master ichnologists. This interpretation however is not supported by Prof. Lockley, who believes that these tracks are of mammalian origin, or made by a still unknown trackmaker (?). A discussion of these two slabs can be found in Lockley and Hunt's 'Dinosaur Tracks', 1995, under the Jurassic Chapter, and in F. Barnes' 'Dinosaur tracks and Trackers', 1997.

Top-of-the-world site (also known as the 'TOW' site). 1972 or 1973 (?)
Upper Navajo Sandstone - upper Early Jurassic, approx. 186 mya.
Discoverer: Lin Ottinger (as per the earlier site)

This large site, made of several slabs still in situ, was then (and still today) one of the rarest and major discovery of mammal tracks ever made in the Southwest. Again, this site was first documented by Prof. Jensen and Stokes (sometime in the mid-80s), then by Prof. Lockley in 1990

-1993. And again, similarly to the earlier 'Two Slabs' discovery, Jensen & Stokes interpreted some of these tracks as *Pteraichnus* (Pterosaur-made) but did agree with Lockley that most of them were *Brasilichnium*, of mammalian origin.

In spite of its importance, this site was, to my knowledge, never fully documented nor for that matter fully investigated (as we shall see later in this record). However, due to Prof. Lockley's reputation, I felt that his documentation and interpretation of these tracks as *Brasilichnium* did not need any further investigation. Thus my earlier visits to this distant site were mostly for bettering my knowledge of this type of tracks. Here is a view of these *Brasilichnium* tracks as then documented:

Most of them are well-imprinted trackway crossing the fairly large and spectacular main slab. All tracks are 'classic' *Brasilichnium*. Made by broad-footed trackmakers because the oval-shaped foot is wider than long (6). These tracks vary in size from approx. 2 - 2 1/2 to 3 1/2 cm (in width), a few showing up to four claw marks. Similar tracks are also present on two other slabs but most of them eroded and lacking clear trackway patterns. All slabs are covered with eroding desert varnish, the main one being the best preserved. According to Jensen & Stokes, and later Lockley, these tracks were made in-around the wet shores of a *playa* (ephemeral and shallow small lakes commonly found in the Navajo Sandstone). Please note the playa interpretation, also invoked at the earlier 'Two Slabs' site. This interpretation is of importance since it was needed to support the potential presence of small crocodilians (via the presumed or real *Batrachopus* tracks mentioned earlier).

From 2000 onward

Breaking-wave site. 2000
Upper Navajo Sandstone - upper Early Jurassic, approx. 186 mya
Discoverer: the author

Located in a remote and difficult to get to area it is the first of a series of mammal sites uncovered since 1972-73. The name stems from its resemblance to an incoming wave actually a horizontal and wetted surface up-lifted later on by the chaotic pressures that have distorted large segments of the Navajo Sandstone. In numbers of track it is by far the most prolific of any mammal sites uncovered to date in-around this region and most likely beyond. Simply put this approx. 10 x 8 ft. site is literally trampled by multitudes of small *Brasilichnium* tracks crossing each other in all directions. Outside these tracks, and near the bottom of the site, is the trackway of a very small Theropod. These multitudes of similar or identical tracks have precluded me to even attempt to make a graphic survey of the site. In that case photographs are vastly more educational than any graphics (7). Here is a brief technical description of these tracks:

All of them *Brasilichnium*-type, rounded or slightly oval, most of them clearly imprinted but showing signs of erosion ('dish-like' tracks), with no clear claw impressions. One continuous trackway made of 17 tracks is clearly visible. This trackway doesn't show any signs of manus impressions. Most of these tracks show a similar diameter (or width) around 2 cm (3/4 in or so) although a few are slightly smaller or larger. These tracks appear to have been made in a wet dune with no traces of a playa in-around the site. Their multitude strongly suggests close-by burrows - but none have been searched for, nor uncovered (so far).

Located approximately 50 yards south of that site, is a well-imprinted trackway of a much larger trackmaker. This trackway is at the same stratigraphic level as the main site, suggesting not only a similar dating but some kind of relationship between these small and large animals - or the opposite, suggesting hostility or perhaps predation from the larger ones. This isolated trackway is made of three tracks: the first and the third showing clear claw impressions (suggesting that the shallower second one may be a manus impression?). The two larger ones have a width of 7.5 cm (3 in) and are the largest mammal tracks ever uncovered in that region, at that time. Due to their morphology - and in accordance with the current interpretation of these tracks elsewhere - I have tentatively documented them as *Brasilichnium* (8). Due to their size and morphology, the trackmaker appears to be an early mammal similar to *Kayentatherium* (9).This tentative association is backed by the body fossil of this animal (its feet structure being well-preserved), but also by its body size incompatible with the diameters of the mammal burrows uncovered in the Kayenta and Navajo deposits (as discussed later).

Culvert Canyon site. 2001
Upper Kayenta Formation - middle Early Jurassic, approx. 191 mya
Discoverer: the author

A mammal trackway located on a vertical 6x6 foot slab that broke-off from a slightly higher bench associated with the Kayenta side of the interface with the upper Navajo Sandstone. 14 continuous footprints in a poor state of preservation, only preserved as rounded 'dishes'. However, the trackway is complete and highly recognizable as mammal-made. The trackway is 27 in. long and 2 in. (4 cm) wide indicating a narrow-bodied animal. Footprints are approx. 2 cm (3/4 ") in diameter. No indications of claws are present. Due to their position both manus and pes impressions are clearly incorporated in the trackway. Using this interpretation strides are steady through the entire trackway with measurements between 10 and 11 cm (4 and 4 3/8 of an inch). In accordance with similar tracks documented elsewhere these tracks are currently interpreted as *Brasilichnium*. These are the first mammal tracks ever uncovered in the Kayenta Formation, still to date.

Pitty-poo site. 2001
Upper Navajo Sandstone - upper Early Jurassic, approx. 186 mya
Discoverer: the author

A 'classic' *Brasilichnium* trackway made of 13 continuous tracks. This trackway is on small slab recently fallen from underneath a ledge. All tracks have a sand-crest impression in front suggesting that the trackmaker was proceeding down a wetted dune. All tracks have a slightly oval configuration and measure 2.5 cm (1 in) in width. The trackway is extremely narrow with no signs of manus impressions - something that should be at least suggested in this very smooth lithified surface. No signs of claws. Based on a manus-pes incorporation the average stride is 18 cm (7 in.). The site is part of a cross-bedded dune with no traces of playa in-around.

The importance of that site goes beyond the fairly common *Brasilichnium* tracks. It is the first one ever uncovered near a (huge) mammal burrow. Near, but not inside. The cross-bedded dunes in which it lays are immediately above the stratigraphic level of this burrow. Nevertheless, when first uncovered, it was the first indication that mammal tracks, particularly the standard *Brasilichnium-*

type, were linked to a fossorial lifestyle, i.e. colonial burrows. This discovery triggered further investigation of that fairly remote area, and subsequently, starting with the next site, three more mammal sites were uncovered in-around this huge burrow; hereby connecting mammal tracks with the multitude of Navajo burrows.

Squirrel tracks site. 2001
Upper Navajo Sandstone - upper Early Jurassic, approx. 186 mya
Discoverer: Terby Barnes, wife of Francis 'Fran' Barnes, of Moab, Utah

The first and most significant mammal site ever uncovered up to that date in the Moab region, and to my knowledge anywhere in the western United States (and possibly beyond). The first tracks that differ - and by a wide margin - from the then dominant *Brasilichnium* types.

This site consists of three small slabs lying next to each other on top of a dome, and totally separated from any surrounding geologic features. Their surface is covered with thick black desert-varnish that preserved these tracks from million years of harsh erosion. So well preserved that details of these tiny footprints are visible 6 foot away. Three trackways, one on each slab. Two in 'elite' state of preservation measuring 13 and 14 in. in length (38 - 40 cm). The other 30 in. long (55 cm) but only in a fair to poor state of preservation. The two 'elite' ones have very small but superbly preserved manus (front foot) imprints.

All hind foot imprints (pes) measure 1 cm in width, and vary in length between 8 mm to 1 cm (.3 and .4 in). All have forward-looking four long digits very close to each other that indeed resemble contemporary squirrel tracks (or similar rodents). No fifth fingers are apparent even under a magnifying glass. In the first 'elite' trackway the stride is 6.5 to 7.5 cm maximum (2.5 to 3 in.). In the second one the stride is 9 cm (3.5 in) on the average. Width of trackways (taken from the stride line of left and right hind feet) is constant around .9 in. (2.3 cm), giving a deflection angle of 30 degree, also constant, between the center of left and right hind feet impressions. This substantial deflection (sometime called pace angulation) suggests a fairly wide-bodied low-slung critter with relatively short legs, typical of contemporary small burrowers. Using the well-imprinted manus and pes impressions the average trunk size (shoulder - hip length) is estimated at around 5 cm (1.8 - 2 in). With a guesstimated 2.5 cm head and an oversized tail at around 8 cm (3 in) in length, the total length of these early mammals must have been around 13-18 cm (5 to 7 in).

Also present at the site are five well preserved trackways of an even smaller animal. The tracks themselves are extremely small, rounded or roundish in configuration with no manus impressions or any sign of claws. They resemble miniature *Laoporus* tracks and like this type of track show only a small deflection from the line of march. The largest (at the very most) measure 5 mm (.2 in) in diameter while most are around 3mm (.1 - .15 in). Assuming that these extremely narrow trackways represent 'all-four' tracks impressions (no manus imprints) they show a constant stride ranging from 3.5 cm (1.4 in) for the somewhat larger tracks to around 2.5 cm (1 in) for the smaller ones. While the trackmakers could be infants of the larger 'Squirrel' tracks maker their origin is still a mystery (10).

One of the slabs also contains a handful of 'dots' tracks. These types of track were also but later uncovered at the Court House site (Wingate Sandstone-very Early Jurassic), then later again at two other sites in the Upper Navajo. It is unclear at this stage what these tracks actually represent, although their presence among documented mammal tracks suggests a mammalian origin (11).

None of these above tracks could have possibly been made and preserved in wet sand, even under the most imaginative conditions. In the Navajo deposit sand grains range in size between .06 mm and 2 mm, too large to have preserved the details of such minute tracks. They must have been made in some kind of extremely fine (and wet) deposit of aeolian origin (dust). This aspect would be of little interest if not repeated in many tracks sites from the Navajo Sandstone, suggesting pluvial occurrences far greater than previously known (12).

This is the first site in this region - perhaps anywhere? - connecting mammal tracks to burrows. The three slabs are inside a huge burrow (the Dewey Bridge Burrow), sitting a very short distance from petrified tunnels. The Pitty-poo site sits at the eastern boundary of this huge burrow, roughly ¼ mile from the Squirrel one, the later being the second one uncovered in-around this burrow (there are 2 more).

As per this date the 'Squirrel' tracks have yet to receive a formal ICZN classification. According to Prof. Lockley, who saw these tracks in the Spring of 2004, the reason could be that similar tracks may already exist in Argentina (13). The tiny ones still do not have even a nickname to identify them.

Mile 11 site. 2001
Upper Navajo Sandstone - upper Early Jurassic, approx. 186 mya
Discoverer: the author

This site is quite similar in size to the 'TOW site' described earlier. It consists of two elevated mounts, one covered with a vast amount desert varnish - the main site - the other, located 50 yards away, quite crumbly and of far less importance.

The main mount consists of several slabs covered with desert varnish. They contain dozens of *Brasilichnium* tracks, most of them badly eroded, or poorly imprinted in what must have been a very wet surface. The better preserved ones are clearly oval in configuration, showing clear, or remnants of, claw impressions, and range in width from 3 cm (1 1/4 in.) to 4.5 cm (1 3/4 in.). No signs of clear trackways, nor signs of any other types of track. Broken pieces of a thin horizon containing extremely rare *casts* of these tracks are lying next to and below one of these slabs (14).

The second mount contains the eroded trackway of a medium-large mammal-like animal. Tracks are of the *Brasilichnium* configuration. This trackway is of importance however, as the tracks change configuration as the animal crosses a small but extremely wet section. On the humid-wet sand part, the tracks are of the *Brasilichnium* configuration, but in the extremely wet section they show-up as rounded, flat 'blobs', obviously made by the feet of the animal sucking mud upward, filling the tracks as it crossed that section. As described in a later site, these rounded flat 'blobs' have permitted permitted identification of the trackmaker as mammalian.

In the vicinity of this site in-around a 100 yards radius there are several biodisturbed areas, some of them covered with remnants of desert varnish. In one of them I uncovered a sole but well-preserved track similar if not identical to the larger type earlier uncovered (2000) in the immediate vicinity of the Breaking wave site. Likewise that track has clear claw imprints and a width of 7.5 cm (3 in.).

Wingate Trail site. 2002
Lower Kayenta Formation - middle Early Jurassic, approx. 196 mya
Discoverer: the author

This extremely remote site located at the interface with the Wingate Sandstone, consists of 3 slabs partially covered with desert varnish. On them are 14 scattered and medium-sized *Brasilichnium* tracks most of the same width and averaging 5 cm (2 in). 3 others of similar configuration are considerably larger with a width around 7.5 cm (3 in). Some are well-preserved but most in various stages of obliteration. Claw imprints are clearly visible in some of them along with sand crescents, suggesting a fairly steep slope. A trackway is suggested among the smaller ones; but this is questionable, as parts of it have eroded away. All these tracks were clearly made in a very wet surface.

Beyond these tracks are footprints of a much larger animal. 4 eroded tracks that appear to be in a trackway (?). These tracks, clearly, are not of the *Brasilichnium* configuration but appear to have been made by some kind of fairly large 'mammal-like reptile' (?). The largest and better-preserved one - most likely a hind foot - is clearly rounded with 5 long claws impressions and a width around 12 cm (5 in). The others are smaller and in poor state of preservation but also show signs of claw impressions (15). These tracks are still an enigma. Further investigation of that very remote site should help trace their origin as mammalian or reptilian.

Court House site. 2002
Lower Wingate Sandstone - very Early Jurassic, approx. 202 mya
Discoverer: the author
Documentation by the author, then partially by Prof. Lockley

The site is approximately 30 yards (100 ft. +/-) in length and close to the bottom of a Wingate Sandstone cliff. An eroding sand-dune with numbers of steeply-angled cross-bedded horizons. The tracks are located in the exposed areas of these horizons The site is made of 3 distinct segments, each separated by what appears to be relatively brief aeolian periods (?).

The first segment contains:

1. A small Theropod trackway made of 4 consecutive footprints in very poor but recognizable condition. No recognizable manus impressions. Distances between footprints are almost identical at around 11 in. (28 cm). Track sizes approx. 1 1/4 in. (3.5 cm) in width and 1 3/4 in. (5 cm) in length.
2. A *Brasilichnium* trackway crossing the Theropod one at a 45 degree angle, and consisting of 6 consecutive footprints. No recognizable manus impressions. Distance between foot prints: 16 cm (6 1/2 in) - except for one at 14 cm (5 1/2 in). These 6 footprints are followed by 3 identical ones, but separated by a small section of broken ground, for a total of 9 footprints. Most of these tracks

are in eroding desert varnish, relatively well-preserved, and on the average measure 2 cm (3/4 in) in diameter. Some of them have clearly visible sand crescents. The 5th track has 2, perhaps 3 ?, rounded toe imprints.

Three very small areas in the immediate vicinity of the above contain various numbers of eroded to fairly-preserved 'Squirrel tracks' varying in size but identical in configuration to the ones earlier discovered at the 'Squirrel site' in the Upper Navajo Sandstone (2001). These tracks (most likely pes) have a width of 1 cm +/-. What appears to be two manus imprints measure 3 and 4 mm in width. These tracks are extremely difficult to detect, especially in glaring sunlight. During one of my many visits I wasn't able to locate them although I knew their general location.

Among the above are a few poorly imprinted tracks that do not appear to be connected to the 'Squirrel' trackmakers. They are represented by 'dots', the only preserved parts of these tracks presumably left by the tip of each claw during progression. These 'dots' tracks have a fairly wide splay of the digits as opposed to the close-together claws typical of the 'Squirrel' type. All the well-preserved ones found at this site (but also in the Upper Navajo Sandstone), have only four 'dot' impressions. Measured across, most have a width of 2 cm or less. What they actually represent is still a mystery. However, they appear to be of vertebrate origin, possibly a badly eroded version (or a poorly imprinted version) of *Brasilichnium* tracks. In the Paleozoic period (Permian) somewhat similar tracks have been identified by Prof. Martin Lockley as *Laoporus* (an earlier version of *Brasilichnium*). In some of these trackways (Lyons Sandstone and Coconino Sandstone) *Laoporus* tracks are present in various state of preservation, from toe impressions to full tracks and oval depressions. The toe impressions do resemble these 'dot' tracks.

The second segment is a 5 x 6 ft. exposed cross-bedding, very fragile and in a state of erosional disintegration (no desert varnish cover). Scattered on that small area are:

Five 'dots' types - 1.5 to 1.8 cm in width. A barely visible trackway made of 3 tiny rounded tracks identical to the ones earlier discovered at the 'Squirrel site' in the Upper Navajo Sandstone (of still unknown origin). Scattered around these relatively well-preserved tracks are several badly eroded track impressions impossible to assign to either 'dots' or 'Squirrel' types. In addition there is a trackway made of 4 consecutive rounded footprints (raised types) averaging 1.2 cm (1/2 in) in diameter, one with a clearly visible sand crescent. These very small rounded footprints have no claw impressions, and appear to be tracks left by an infant *Brasilichnium* trackmaker (?). This tentative interpretation is solely based on the rounded configuration of the tracks.

The very small third segment (8 x 12 inches) is located on a protruding cross-bedding, and like the second one is extremely fragile and in a state of erosional disintegration (no desert varnish either). It contains 3 to 4 very small and badly eroded tracks that can be assigned to either 'dots' or 'Squirrel'. It also includes the smallest tracks ever found of what could be a tiny mammal-like animal (?). These two tiny tracks have a narrow triangular shape with elongated digits (claws? digits?). The larger one, possibly a hind foot impression, is 5 mm in width, the other 3 mm. The smaller one, due to its location next to and slightly ahead of the larger one, appears to be a front foot impression. These two tracks are almost impossible to detect to the naked eye but clearly stand-out under a magnifying glass.

actual size

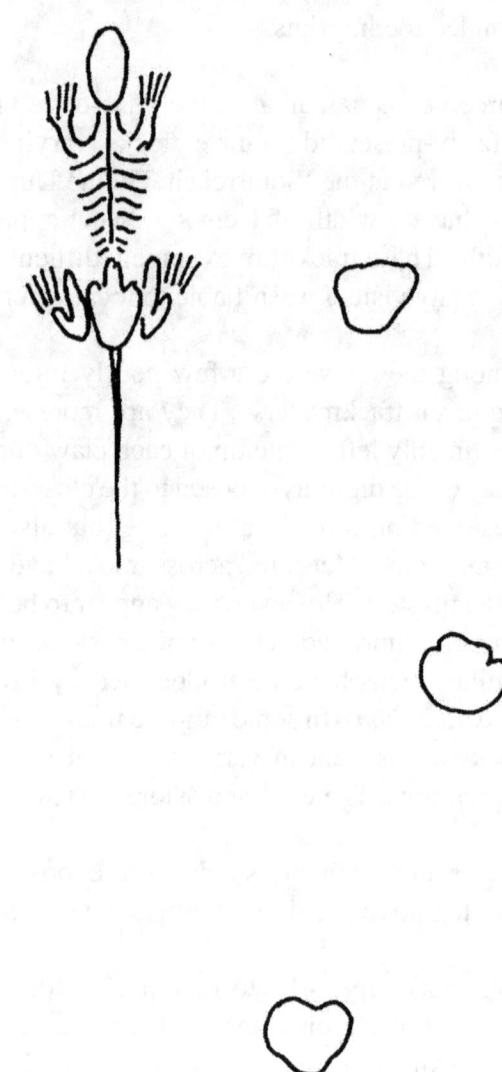

A burrow entrance with scattered tracks possibly made by a diminutive early mammal. Eroded trackway on the right is of the Brasilichnium confi guration, with no manus impressions although made in a soft wet surface. A life-size schematic rendering of the Early Jurassic Hadrocodium was added for scale. Top-of-the-World site, Upper Navajo Sandstone – end of Early Jurassic, approx. 186 mya.

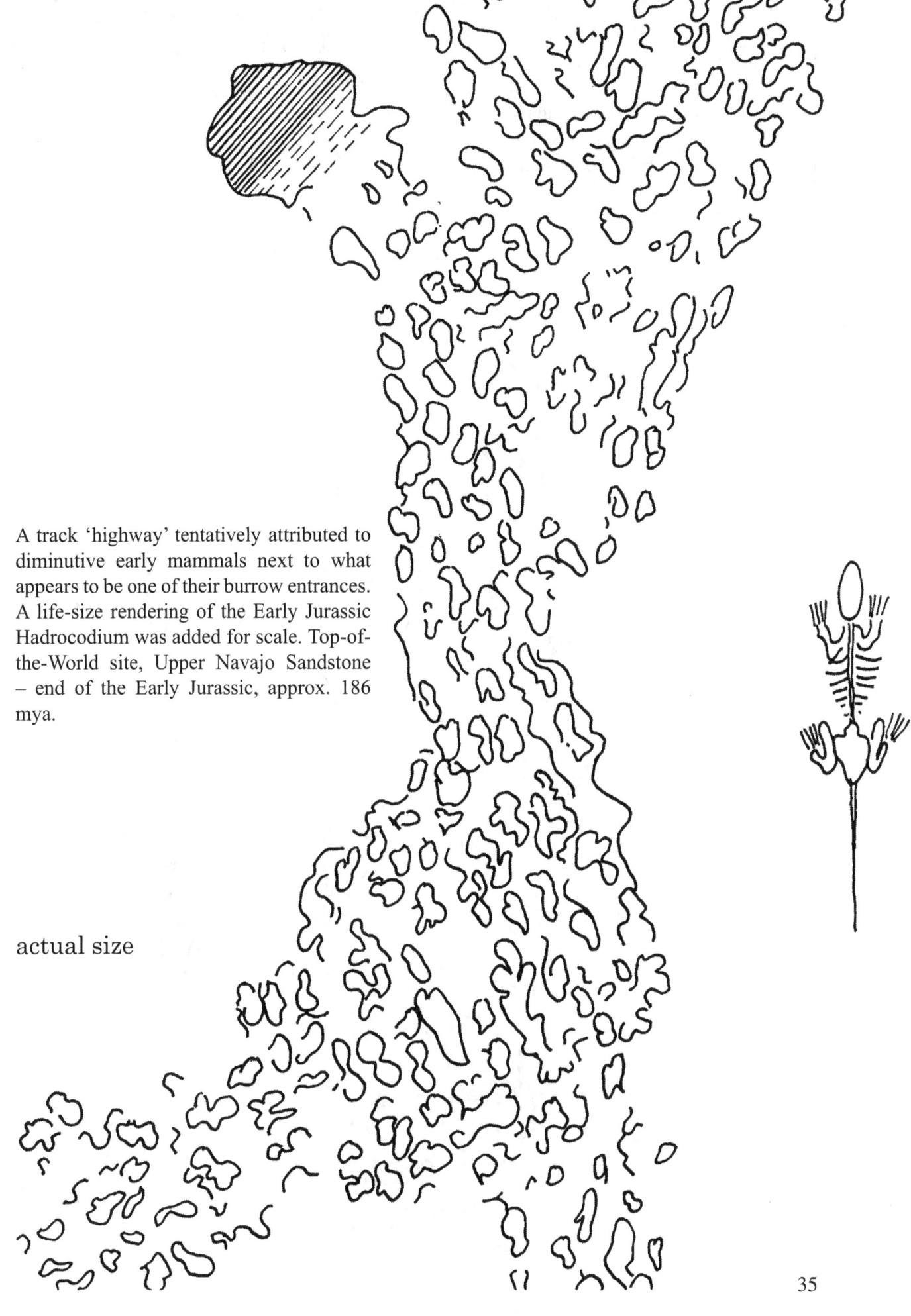

A track 'highway' tentatively attributed to diminutive early mammals next to what appears to be one of their burrow entrances. A life-size rendering of the Early Jurassic Hadrocodium was added for scale. Top-of-the-World site, Upper Navajo Sandstone – end of the Early Jurassic, approx. 186 mya.

actual size

Highways' located in-around burrowed areas that appear to have been made by very small mammals although the possibility exists that they could also have been made by some kind of invertebrates. Upper Moab-Entrada deposit – end of the Middle Jurassic, approx. 165 mya.

actual size

"Squirrel" type of tracks made by an unknown but very small burrower. Notice clear manus imprints. Part of three superbly preserved trackways. Similar tracks were later uncovered in the very Early Jurassic (lower Wingate Sandstone). Upper Navajo Sandstone – end of the Early Jurassic, approx. 186 mya.

actual size

Trackways made by either tiny mammals or unknown invertebrates. Their presence next to "Squirrel" trackways seems to indicate a vertebrate origin, possibly infants. "Squirrel" site, Upper Navajo Sandstone – end of the Early Jurassic, approx. 186 mya.

actual size

Part of a typical Brasilichnium trackway. Semi-circles in front of footprints are sand crescents. Pitty-Poo site, Upper Navajo Sandstone – end of the Early Jurassic, approx. 186 mya.

actual size

actual size

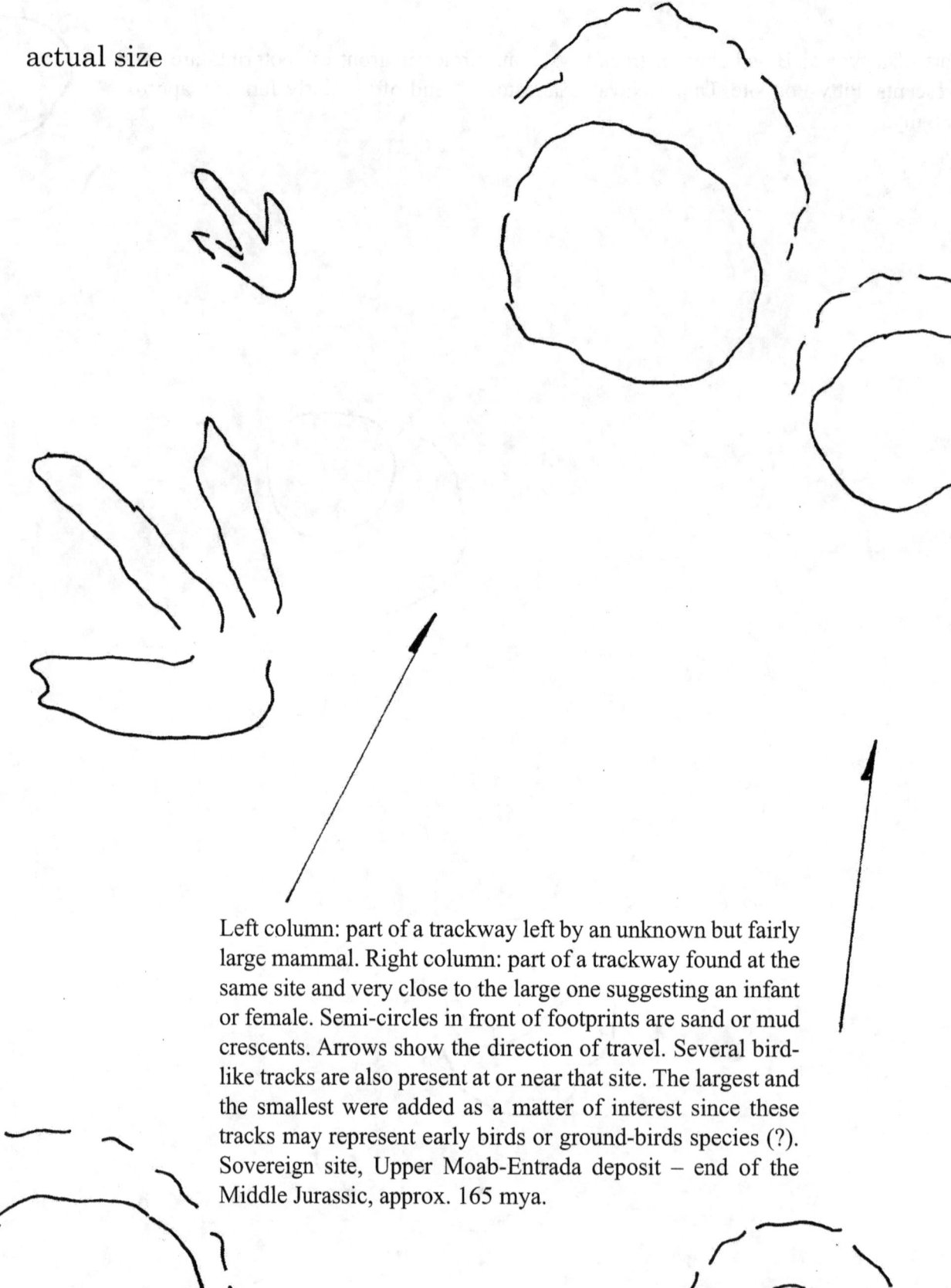

Left column: part of a trackway left by an unknown but fairly large mammal. Right column: part of a trackway found at the same site and very close to the large one suggesting an infant or female. Semi-circles in front of footprints are sand or mud crescents. Arrows show the direction of travel. Several bird-like tracks are also present at or near that site. The largest and the smallest were added as a matter of interest since these tracks may represent early birds or ground-birds species (?). Sovereign site, Upper Moab-Entrada deposit – end of the Middle Jurassic, approx. 165 mya.

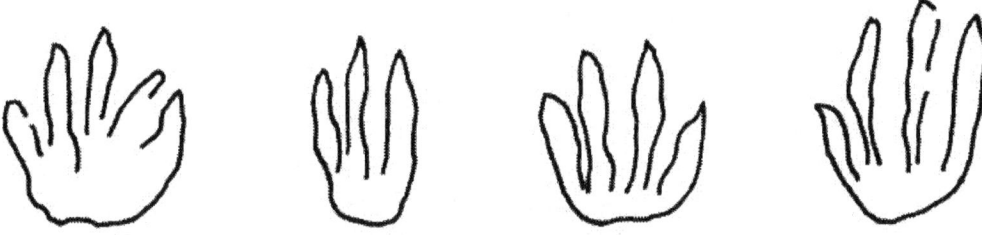

"Raccoon" type of tracks. Notice extremely elongated digits and near-total absence of the ball of the foot. The four digits ones appear to be hind feet impressions, with the three digits one possibly a manus one (?). Raccoon site, Upper Navajo Sandstone – end of the Early Jurassic, approx. 186 mya.

'Dot' tracks. They are still a mystery but appear to have been made by some kind of early mammals. This tentative interpretation is based on somewhat similar claw impressions left by Laoporus trackmakers in the Permian Period. Court House site, Lower Wingate Sandstone – very Early Jurassic, approx. 203 mya.

actual size

These above tracks, including the *Brasilichnium* trackway, are the first ever uncovered at the very beginning of the Early Jurassic in southeastern Utah. What they actually represent is unclear, until more are uncovered and studied in following years. However, since they do not fit any known insect or invertebrates tracks like amphibians, lizards, etc, we must at least take into consideration that they, or some, may be of early mammals origin.

Note: This site is the most graphic example of the difficulties faced by scientists and 'trackers' in their search for mammal tracks, and the very reason why they are so few in the track record. This particular area, easy to access, was known to me and investigated on several occasions, with only scattered *Grallator* tracks (early Theropod tracks) recorded. The site itself, with exposed cross-bedded horizons, had indeed attracted my attention, yet I was unable to detect anything, except the larger *Brasilichnium* trackway. Knowing that tracks 'stand-out' as the sun goes down, especially in Winter months, I again checked these horizons by sheer curiosity. And here were these tracks.

Alcove site. 2002
Lower Wingate Sandstone - lower Early Jurassic, approx. 201 mya
Discoverer: the author

A trackway made of 6 badly eroded but consecutive footprints. This trackway appears to have been made on wet sand and a slightly slopping surface. The footprints also indicate a slight tilt to the left of the line of progression. This trackway is accompanied by a parallel one made of 5 imprints so disintegrated (except for one) that they could be easily brushed-off as erosional features. Both are difficult to locate under intense sunlight.

The main trackway measures 52 in. (4 ft. 4 ") in length. Distance between footprints is almost identical (also taken from center to center), as follows: 1st step (going up the slight slope) 28 cm (11 in), 2nd, 25.5 cm (10 in), 3rd, 25.5 cm, 4th, 28 cm, and 5th, 25.5 cm. These measurements indicate that this trackway was made by an animal and not the result of any possible erosional features. Width for the first four footprints (beginning at the top of the gentle slope) measures 9 cm (3 1/2 in) and the last two 2 7 cm (3/4 in). These measurements are approximate due to the ill-defined outline of the footprints but are within that range.

Footprints are rounded, dish-like, the standard product of long exposure to erosion. From the top of the slope down the 2nd and 3rd footprints (including perhaps the 4th?) seems to indicate short claws although they could be products of recent erosion. There are no manus indications, something that should have been present in such wet surface and for such a large animal. This suggests either a quadruped animal walking on 'all-four', or the badly eroded tracks of a bipedal dinosaur. Either way, the narrow trackway indicates a narrow-bodied, fairly tall and fast moving animal (16).

For the record the other parallel trackway is 43 in. in length (again from tracks center), with a stride exactly 11 in. between the 5 imprints, if in fact they are animal-made.

These tracks are the first ones of that type and size ever found in the very Early Jurassic of the Southwest. As per this date they remain a bit of a mystery. It is nearly impossible - due to their eroded state - to associate them with *Brasilichnium* types. However, backed by the up-to-date mammal track record, they may have been made by some kind of early mammal. This tentative interpretation stems from the narrowness of the trackway. Outside a mammal-like animal, only bipedal Theropods (or Ornithopods?) could have left such a narrow trackway but the distance between the footprints doesn't seem to fit any of these well-established dinosaur tracks.

Upper Willow Spring site. 2003
Upper Entrada Sandstone Formation - upper Middle Jurassic, approx. 166 mya
Discoverer: the author

These are the first tracks of anything ever uncovered in this 5 million year deposit known as the 'Death Zone' - the last aeolian period of the Jurassic in this region. They were found along the bank of a very small narrow canyon that cuts through the upper Moab-Entrada deposit and down a few horizons into the Entrada Sandstone. They are located on a very small exposed area of a cross-bedded dune.

Four distinct trackways, plus two highly eroded, but still visible ones, along with a scattering of other but similar tracks through two very thin bedding planes. All tracks are eroded, rounded, and 'dish-like'. Their diameter vary between 1.5 and 2 cm (.6 to .8 of an inch) - except for a short trackway with 1 cm (.4 of an inch) diameters. One short trackway crossing the dune diagonally, clearly shows downward slippage. All trackways are extremely narrow with no signs of manus impressions. Distances between footprints indicate an original steep surface: trackways with shorter distances, averaging 8.5 cm, going up that surface while two others with larger distances averaging 12 cm (including tracks showing slippage) must have been crossing it. This appears to be the only explanation for this difference since the diameter of the tracks is constant.

Note: At the time of their discovery, these tracks raised enough controversy, due to their location in the Entrada Sandstone, that I took the time to re-investigate them on three other occasions. As per this date I have no other explanations to offer than a mammalian origin, more so after the subsequent discovery of mammal burrows in the upper Moab-Entrada Member of the Entrada Sandstone. Small, rounded 'dishes' of geologic origin are relatively common in aeolian deposits, and without closer inspection can be construed as mammal-made. Such geologic dishes are however scattered around with no or any suggestion of trackways. In Ichnology, three consecutive impressions at similar intervals, and in the same direction, usually indicate animal tracks. The case at this site.

Raccoon site. 2003
Upper Navajo Sandstone - upper Early Jurassic, approx. 186 mya
Discoverer: the author

This very remote and extremely difficult to detect site consists of two very small slabs lying flat on the ground and next to each other in a dry, shallow, and rocky wash. These two horizontal slabs are broken but in situ (part of the original surface). Recent rain water had washed dirt and debris from their surface permitting their detection but debris may again be present making these slabs very difficult to locate. No signs of playas are present in-around the site. However, an investigation of the surroundings indicates they were part of a fairly large, horizontal, and wet horizon; horizons that until recently, were unknown or poorly documented in the Navajo Sandstone (17). Further yet, this horizon appears to be 'burrow-disturbed', albeit in very poor condition. Later and more extensive investigations should confirm or not this burrowed horizon (a Class III type, as explained in the next chapter).

Note: In December 2005, two *de-watering pipes* were uncovered nearby and at the same geologic level, confirming the earlier extremely wet horizon. These pipes also suggest an eroded but 'burrow-disturbed' area today documented at other sites. De-watering pipes are explained in a later chapter.

This site contains nine small tracks of a type undocumented in the track record anywhere (to my knowledge). Seven are on the small light-colored slab, the two others on the next upper and darker-colored one, broken but connected with the first. They are somewhat scattered with no visible trackway. These tracks resemble contemporary raccoon footprints - thus the name for the site and their temporary nicknaming as 'Raccoon tracks'.

These tracks have extremely elongated and narrow digits with very small heel impressions. All have four digits (except a smaller one with only three) measuring 3 cm (1.2 in.) in length for the largest to 2.2 cm (0.9 in.) for the smallest. Their width ranges between 1.7 to 2.4 cm (0.7 to 0.9 in.). Pending comparative studies of contemporary mammals, somewhat similar to this ancient trackmaker, it appears that these extremely elongated digits are not claw impressions, only digits ending with fairly small claws. This tentative interpretation is based on irregularities or distortions of the digits in some of these tracks along with some sharp extremities suggesting some kind of claw.

The closest resemblance to any known tracks is *Rhynchosauroides*, a Triassic track attributed to an earlier small lizard. This particular track was very common in the Triassic but rare in the Jurassic (18). This vague resemblance is only mentioned because there are no other tracks even close to this 'raccoon' configuration. However, to attribute this new track to some kind of lizard would demand 'stretching' their digits (if not the imagination). On the contrary, these tracks appear to be within a 'burrow disturbed' area fairly close to the Mile 11 site with its dozens of *Brasilichnium* tracks, suggesting a mammal connection. Further, with the recent discoveries of the 'Squirrel tracks', *Repenomamus*, but more important, *Fruitafossor* in the Late Jurassic of this region, to arbitrarily attribute these new tracks to lizards or other invertebrates would be another bias against the presence of multitudes of early mammals now confirmed by their gigantic burrows. At time of publication these 'Raccoon' tracks have yet to be seen 'in situ' by anyone outside myself.

Squirrel site II. 2003
Upper Navajo Sandstone - upper Early Jurassic, approx. 186 mya
Discoverer: the author

NOTE: That site was buried by a road-grader in May 2005. However, due to later heavy rains, dirt and gravel should wash away from its solid rock surface.

This site is located in the vicinity of the above Raccoon site (approx. 200 yds. away). It appears to be at the same stratigraphic level and also within a large wetted area (perhaps the same one?). The site itself is a 'strip' of hard flattish rock approx. 4 feet wide and 30 feet long, parallel to a graded county road, and recently exposed by surface water action.

Scattered along this 30 foot 'strip' is a small number of 'dots' tracks, some well-preserved, some badly eroded and barely visible. Some smaller but badly eroded ones may be of the 'Squirrel' type (?). The well-preserved 'dots' types are, or appear to be identical, both in size and configuration to the ones first discovered at the earlier Court House site (early Wingate) - and probably identical to the ones also discovered at the 'TOW' Site during a recent re-investigation.

This site would be of marginal interest if it wasn't located near the Raccoon site and the Mile 11 one. That general area, located 11-12 miles on a dirt county road connecting with a paved highway 30 miles from Moab, Utah, is extremely rich in geologic features, and thus of importance in spite of its remoteness.

Top-of-the-world site. Re-investigation, 2003
Upper Navajo Sandstone - upper Early Jurassic, approx. 186 mya
Made by the author

The following results are incomplete. Due to the restricted time factor imposed by its remote location, that re-investigation was mostly focused on types of mammal tracks potentially present on the various slabs lying around that site. More thorough re-investigations will be scheduled as time permits. In the mean time the following already represents a vastly different view of that site - more important, it introduces an entirely new type of tracks now known as '*highways* '.

Beyond the already documented *Brasilichnium* trackways the following track-types were uncovered during this preliminary re-investigation:

1. 'Dots' tracks, similar or identical in size and configuration to the ones first documented in the Lower Wingate (Court House site) then in the Upper Navajo Sandstone (Squirrel site II). These tracks, a handful well-preserved but most in various stages of deterioration and barely visible, were uncovered on two smaller slabs lying 50 ft. from the main one. They are scattered with no indications of trackways.

2. Among the above 'dots tracks' are a handful of poorly preserved 'Squirrel tracks'. Or, to be exact, what appears to be 'Squirrel tracks'.

3. Two trackways of a very small animal. This type of trackway was first documented at the 'Squirrel Site" then found again at the 'Court House site ' (lower Wingate). As already described in the other sites these trackways are extremely narrow, indicating not only a narrow-bodied animal but a fairly long-legged one as well. At this site the diameters for the rounded footprints are 3-4 mm for the smallest and 4-5 mm for the largest. Distances between footprints vary between 2.5 and 3 cm (1 and 1.2 in.) for the smallest, and no more than 3.5 cm (1.4 in.) for the largest. These measurements are comparable with the trackways found at the other sites. As mentioned earlier, the identity of the trackmaker is unknown but its presence (again) among documented mammalian tracks seems to suggest infants or some kind of tiny mammals.

4. Two small trackways made by either infants *Brasilichnium* or very small mammals. These two

trackways are very similar if not identical. All tracks are rounded with no signs of claws (most likely due to the original murky surface), with their widths constant at around 1 cm (3/8 of an inch). However, these trackways are not narrow (as per 'classic' *Brasilichnium* ones), and comparatively show a fairly wide pace angulation. Track width is similar to the 'Squirrel' trackways documented at the 'Squirrel' site. This observation, along with a constant distance between footprints, 5 cm (2 in.), seems to indicate small mammals larger than 'Squirrel' trackmakers, but similar in anatomy, thus possibly associated with the 'dots' trackmakers (keeping in mind that while 'dots' tracks are relatively numerous none of their trackways have been uncovered to date). Please note that these interpretations are tentative pending further investigation.

5.   Three '*trackways-highways*', made by either tiny mouse-like mammals or by similarly sized insects, connecting four very small burrow entrances. To describe such '*highways*' the accompanying photograph and graphics are a lot better than words. In a condensed version they are paths or trails made by tiny animals traveling back and forth from their burrow entrances. In so doing these trails - or 'highways' - have been trampled to a level where individual tracks are nearly impossible to detect. As a point of reference, to better grasp the context and size of these 'highways', I added a rough (but accurate) drawing of the *Hadrocodium* fossil - the smallest early mammal ever uncovered to date in the Early Jurassic.

In addition, this re-investigation did not uncover any signs of *playas* in-around the site, as suggested or claimed in the original investigation(s). The only playa around is located along the dirt county road (300 yards away) and well-below the stratigraphic level of the site (although severe up-lifting in that area may have affected the original level of the site). A thorough investigation of that playa (made in 2002) did not yield any tracks. Please note that *no mammal tracks of any kind have ever been uncovered in playas in-around the Moab Region* (or anywhere I know of), in spite of fairly thorough investigations during the past 15 years.

The original claim that the site was inside or next to a playa must be seen through the very limited knowledge of the Navajo Sandstone at that time. In other words, all the tracks at that site being made in a wet surface, they must have made inside or on the shores of a playa since these ephemeral lakes were the only known body or water present at that time in the Navajo deposit. Today, with hundreds of wet horizons, springs, etc., being continually documented in that deposit, playa determinations are no longer valid. However, it should be noted that among the original scientists who investigated that site ,four of them were geologists with Phds. Yet none of them saw, nor were intrigued, by the heavily biodisturbed (wet) horizons lying immediately above the site. This shows the dogmatism that was still present at that time in both geology and paleontology (still present unfortunately in many of these quarters). The Navajo Sandstone was then nothing more than a desolate desert, dotted here and there with ephemeral playas, and inhabited by a few Theropods 'eating each others'. The wet and biodisturbed horizons on top of that butte are so blatant and intriguing that is impossible for any professionals to miss them. Yet they did. No wonder they also missed the other tracks at that site, including 5 casts of an unknown but very large animal located only 15 feet behind the main slab containing the best preserved *Brasilichnium* trackways. By today' standard these were sloppy investigations.

Burrows (Class II, originally uncovered in 1999 near the bottom of the butte where this site is located), were extended to the area east and south of it, but as eroded Class III types. Time did not permit to extend the investigation of burrows beyond these two limited areas. However, these burrows infer that the mammalian trackmakers at that site were most likely colonial and fossorial, at least partly.

Sovereign site. 2003
Moab-Entrada tongue of the Entrada Sandstone
Upper Middle Jurassic, approx. 166 mya.
Discoverer: the author

Note: Actually 5 sites containing similar or identical tracks. The best preserved and most important is by far the Sovereign site, and an adjacent but poorly preserved one encompassed under the same name. The other three are also at the same stratigraphic level as the Sovereign one but located up to 5 miles east of it. One is in a non-descript area approximately 3 miles away, the other is east of the Klondike Trail, the last one also containing some small tridactyl tracks is located approximately 100 yards south of the famed Megatracksite complex uncovered and documented by Prof. Lockley. Geologically, all these sites are at, or close to the interface with the upper Summerville Formation, a local formation considered by many geologists to be the beginning of the Late Jurassic in southeastern Utah.

The Sovereign site is made of four segments, three very close to each other - the main site - and the fourth approximately 150 yards to the north. All of them are located on fairly narrow strips of exposed cross-bedding. The three segments of the main site are on different cross-bedding but very close stratigraphically, while the fourth is impossible to tell due a Summerville mount separating it from the others.

All tracks (except tridactyl ones) uncovered so far in these various sites are rounded or slightly oval with no claw or manus impressions. They only differ in size: 5 cm (2 in) in diameter for the largest, the most numerous, and likely made by mature animals. 3 cm (1 2/8 in) in diameter for the smallest - probably made by infants or females. Some have sand crescents in front indicating a gentle slopping surface, but others do not. Also, outside a few scattered ones, most of them are in trackways.

At the main site the best preserved trackway is extremely narrow (like all others found to date) and consists of 6 (perhaps 7?) well-imprinted tracks of the larger diameter (5 cm -2 in). Distance between footprints is constant between 18 to 20 cm (7 and 8 in). Other trackways are lesser preserved but clearly visible, all with similar diameters. Also present at that site are the smallest of that configuration found so far. They are in a trackway next to larger tracks. 5 tracks measuring on the average 3 cm (1 2/8 in) in diameter. Scattered around are a few tracks of both the larger and smaller diameter.

All these tracks were made in very wet surfaces. In such surfaces smaller front foot (manus) impressions could not have been preserved, and there are none. However, this is to presume that the animal had smaller front feet, today a notion that has been dispelled in many Mesozoic and contemporary tracks made by similar animals. Either way, we are dealing here with a fairly large animal. The trackways are narrow, similar to predatory species like contemporary coyotes, indicating tall and most likely fast moving species. Further, and of great importance to its identity, no burrows so far uncovered in this deposit (or anywhere in the Navajo Sandstone), could have housed such a large animal (somewhat similar in size to the Early Jurassic *Kayentatherium*).

Determination of these tracks as mammal-made, whatever the classification, was an exercise in both dogmatism and bias against mammals. As per this date, the most likely track type is *Brasilichnium*. This is where the *Brasilichnium* trackway crossing a very wet section at the Mile 11 Navajo site comes into the picture (please refer to it). These Sovereign tracks are identical or nearly identical, but some are larger. Whatever the future classification of these tracks will be, their size proves beyond discussion that fairly large mammals existed during that late period of the Jurassic in southeastern Utah, perhaps also elsewhere ?. The real value of this discovery (19).

In addition, a fairly large numbers of small to very small tridactyl tracks are present in-around these mammal sites. A welcome discovery because until then these bird-like tracks were extremely rare at that geologic level. Most of them range around 3.8 cm (1 1/2 in). in length. The largest measures 7.5 cm (3 in), and the smallest (possibly the smallest ever found anywhere) is barely over 2.5 cm (1 in). These newly-found (and increasing) numbers of very small tridactyl tracks seem to suggest that some kind of early birds or bird-like Theropod species may have been present in the upper Middle Jurassic of this region. These bird-like tracks were documented by Prof. Lockley in May 2004.

The Third site. 2005
Upper Navajo Sandstone – Upper Early Jurassic, approx. 186 mya
Discoverer: the author

Nicknamed the 'Third site' as it is the third one uncovered in-around the huge Dewey Bridge mammal burrow after the 'Pitty-poo' and 'Squirrel' sites. It is located a ¼ mile, at most, from the earlier sites in a small wash along the graded county road. Tracks are in a narrow bedding exposed by recent water erosion. All of them are badly eroded except for four small-medium-sized ones that are of the classic *Brasilichnium* type. This site has little scientific value outside of its presence around the Dewey Bridge burrow, again suggesting numbers of early mammals living at the very end of the Early Jurassic of Utah.

The Fourth site. 2005
Upper Navajo Sandstone – Upper Early Jurassic, approx. 186 mya
Discoverer: the author

Nicknamed the 'Fourth site' as it is the fourth uncovered (so far) in-around the Dewey Bridge burrow. It is located only 100 yards away from the 'Third site', also in a similar wash. This site scientific value is very high however since it is the second one in an excellent state of preservation ever found of the 'Squirrel' type. A straight trackway made of 18 very small dish-like (eroded) tracks clearly representing the hind feet of an animal similar or identical to the 'Squirrel' trackmaker. The stride between tracks, and the pace angulation, are nearly identical to the 'Squirrel' type. The dish-like imprints have a diameter of 1.5 cm, ½ cm larger than the elite tracks at the 'Squirrel' site, but this higher measurement is standard in all dish-like tracks due to erosion. A very small animal, and clearly a mammal.

Note: For the ones interested in searching for mammal tracks this 'Squirrel' trackway was almost impossible to detect due to sun glare, color of the slab, and dirt lying over it. What 'gave it away' were unusual marks on a normally smooth slab, marks that could be seen from 50 feet away although they could have been made by moss, gravel, dirt, or discoloration. In the field any unusual marks demand inspection. This I did, and only 5 feet away, realized I was looking at a potential trackway, as 3-4 tracks were visible at its upper end. Brushing, then wetting with a spray water bottle, showed the entire trackway. Even then, due to the sun glare, this trackway was nearly impossible to photograph. Chalk outline of the tracks was needed to make them stand-out against the surface slightly darkened by the evaporating water spray.

Additional tracks ?

In his last booklet, 'Footprints on the shores of time', Canyon Country Publications, 2003, published after his death in late 2003, Francis 'Fran' Barnes shows some pictures of mammal tracks, with no information about their origin, date, location, discoverer, and subsequent documentation. All, except two, are from my discoveries. Under an arrangement made prior to his writing, I was his teammate and we shared all our discoveries right up to the last two weeks prior to his passing. The only track site I did not investigate was one in the Sand Flats Area of the Navajo deposit, he had mentioned earlier to me. This site, as described verbally to me, is most likely of mammalian origin, although today I have no way to find out where it is located. A second one in the same area turned out to be of insect origin. Barnes' booklet is not about tracks per se, but a personal analysis of the problems faced by ichnology in the determination of fossil footprints. Although I do not share some of the views, this booklet nevertheless points to the same 'one-dimensional' problems that have plagued paleontology and ichnology over the years. I recommend it to anyone interested in field research.

Why so many mammal tracks sites in this relatively small region of Utah ?

Look for them and you'll find them, almost anywhere, and hopefully, these small tracks can be preserved. Prior to 2000, the 'Two-Slabs' and 'Top-of-the-World' sites were the only two major discoveries of mammalian tracks, thus investigated and documented by many high-ranking scientists (20). However, these two large sites were easy to see and detect for anyone experienced in field tracking. Then, for over 20 years, no more mammal tracks. Beginning In 2000, I am the one who began looking for them, again, due to my interest in early mammals. Today, to my knowledge, I am still the only one looking for these tracks in the entire West. To find them demands a lot of time, endless hiking, but more important, a knowledge of regional geology and a 'feel' for mammal habitats during that period, as opposed to 'dinosaur tracks' that are relatively easy to detect due to their size and deep imprints. It also demands a strong interest in early mammals, something shared by only handfuls of scientists, most with careers solely or mainly linked to dinosaur studies. In the Moab region there were no more early mammals than anywhere else in the world, in fact far less. That it took 100 years to locate 2 mammal tracks sites in the region, but only 5 years to locate 15 more is nothing new. Only a matter of interest. Hopefully one of you will follow that path, as the story of the Mesozoic, particularly the Jurassic, no longer rests on overblown dinosaurs but on our ancestors, the early mammals.

What about other tracks and their significance?

Until recently the Navajo Sand-Sea was dominated by only one type of track: *Tridactyl*, also known as 'Dinosaur Tracks', large or small, with three digits resembling bird tracks, a type of track attributed to carnivorous *Theropods* or herbivorous *Ornithopods* – both dinosaurs. These ubiquitous tracks are found across all continents from the Late Triassic to the end of the Cretaceous. Near the end of the Jurassic, some very small but similar tracks are tentatively attributed to early birds, or 'dino-birds'.

By comparison, tracks attributed to other vertebrates (except mammals or mammaloids) are rare to extremely rare. In order of importance, *Otozoum* attributed to *Prosauropods*, *Pteraichnus* attributed to *Pterosaurs* (flying reptiles), *Batrochopus* attributed to crocodilians, and some fairly large tracks, like *Navahopus*, attributed to Prosauropods, crocodilians, or mammaloids, depending to whom to talk to. Please note that tracks attributed to crocodilians have only been found at or near the shores of the Western Sea (the western boundary of the Navajo deposit). Diminutive tracks of small, various invertebrates (ants, scorpions, spiders, salamanders, toads, etc.,) have been uncovered in-around ephemeral lakes, oasis-springs, or wet beddings, but their number is still minimal. Beyond this assemblage, no tracks of crocodiles, turtles, or any other kind of *reptile,* have been uncovered to date anywhere in the *interior* of this gigantic sand-sea.

Based on this up-to-date track record *the two dominant vertebrate species in the Navajo Sand-Sea*, in numbers and by far, were dinosaur *Theropods* (most likely with similar but herbivorous *Ornithopods*), and *mammals-mammaloids.* To assign or suggest any other kind of vertebrate to the recently uncovered and gigantic Navajo burrows would **contradict** this up-to-date track record, unless other evidence is provided to confirm their presence.

Side-bar

'Tracking' vs. Field Research

An evolutionary aspect of Ichnology

In the Comments Chapter of my earlier book, 'The Jurassic, a New Beginning?', under the 'Tracks, Trackers, & Science' sub-title, I briefly discussed the pluses and minuses of field exploration and how they affect our perception of the ancient fauna. Although important at the time, this sub-chapter did not dwell into the differences between 'Tracking' and field research. I will do so in the following essay, as the gap between these two versions of field exploration is growing larger by the year.

Ichnology is the study of tracks. But to study them *they first have to be found*. To date most tracks have been uncovered by hikers, ranchers, outdoor enthusiasts, and other persons enjoying such activities. Some, with a basic knowledge of ichnology, will go beyond the hiking-stumbling part and will go to some length in search of tracks in a given area. To this group, we add a very small contingent of students of paleontology, ichnology, geology, or other related sciences, including some paleontologists-ichnologists carrying out field documentation. This academic group is of course, better equipped than the amateurs in their searches for tracks, but their impact on the overall track record is limited by their usually very short stay in the field. A condensed but fair representation of 'tracking' - The very foundation of Ichnology.

In the Early and Middle Jurassic of southeastern Utah - and by extension most of the Western United States - the fossil record is so poor that the fauna is mostly known (and in some region, only known) from the tracks they left behind. But to date, most of these tracks have been uncovered by amateurs, not by professional or semi-professional field researchers. Thus the track record is not a true representation of the fauna at hand, only a superficial view offered by benevolent but largely uneducated 'trackers'. And when the track record is superficial so is the *fauna*.

This is exactly what happened in southeastern Utah, and by extension the entire Southwest.

In this region the track record is almost completely made of tridactyl 'Theropod tracks', most medium to large. The reason behind this over-abundance of 'Theropod tracks' is simply because they are easy to spot. Thus the scarcity of other types of tracks (particularly the small to very small mammal ones) has little to do with the realities of the Early-Middle Jurassic fauna, it only represents amateur tracking.

In the past and still today, this over-abundance of 'Theropod tracks' has contributed to a conventional and disproportioned view of the Early-Middle Jurassic fauna, since most paleontologists take this incomplete track record at face-value. Today, with increasing challenges from other disciplines the scientific authority of ichnology is at stake. And the only way to ward off challenges and enhance the validity of the track record is to switch from current field amateur standards, to professional, or semi-professional ones.

Here is a summary of the differences between amateur tracking and field research:

1. First, it's literally impossible for anyone - even the most qualified scientists - to thoroughly explore even a medium-sized area without a very long stay, and in many cases (as in the Moab region) a year-around presence. This is practically impossible for amateurs and scientists alike, since most are living away, sometime far-away, from the given area. Further, most competent scientists have sedentary careers, that seriously restrict field explorations. To this we must add that these scientists are highly underpaid for their work, and thus unable to afford costly far-away explorations, unless some financial support is available to them; a highly problematic affair nowadays. Foundations' grants are seldom available for speculative explorations, unless the foundation is certain to benefit from a major discovery. And 'rat-like' early mammals or similarly obscure critters are not part of the program.

2. Second, and most important, the days of easy and major field discoveries are over. Today, to uncover anything of true scientific value, explorations demand ever increasing levels of scientific knowledge, including some statistical and analytical ones. Plus experience only acquired through continual and increasingly educated investigations. This translates into year-round or extended field work backed by pertinent studies. No longer the domain of amateurs.

In summary, the future of ichnology lays not in the hands of benevolent 'trackers', albeit welcome today and into the future. It lies in the hands of professionals or specialists with either an academic background or training in this very specialized field. The discovery of these gigantic burrows and their tracks accompaniment are the living proof of that trend.

Chapter Index

1. In their book 'Dinosaurs Tracks', 1995, Lockley and Hunt offer a comprehensive view of the *Laoporus - Brasilichnium* debate, but more important offers a very detailed view of the then mammal record in the western United States. Their documentation, illustrations, and accompanying photographs are still today the most comprehensive overview of the overall track record in the US. Since a great deal of Lockley's field work took place (and still does) in the Southwest this book is a 'must-read' for anyone interested in ichnology, particularly in this region. Lockley's latest book, 'A Guide to the Fossil Footprints of the World', 2002, a Lockley-Petersen publication, is also a must-read for it gives a wider view of the world track record, albeit in less detail.

2. The contributions made by Spencer Lucas toward a better understanding of the *Laoporus* tracks in the Permian Period are, if not unique, at least of the highest scientific standards. In his 'Early Permian Footprints and Facies', 1995, a book published by the New Mexico Museum of Natural History and Science and co-edited by Andrew B. Heckert, he describes in sometime minute details, both the difficulties of identifying Paleozoic trackmakers and a deep insight into the *Laoporus* origin. Numbers of accompanying photographs are by themselves one of the most inclusive overview of the Permian Period.

3. After Lockley and all, 1993

4. The full story of these two slabs is described in Francis 'Fran' Barnes' 1997 book, ''Dinosaur tracks and Trackers'', Canyon Country Publications

5. The *Batrachopus* interpretation stems from a long debate with Jim Jensen & Lee Stokes on one side supporting a pterosaurian origin (*Pteraichnus*), and K. Padian, P. Olsen, and G. Leonardi supporting a crocodilian one, with Martin Lockley supporting neither of these interpretations, instead a potential mammalian origin. This is not, however, the end of the debate, only the beginning of a newer and more sophisticated phase. Please refer to the later chapter 'The one-dimensional issue'.

6. Lockley's own words and one of the best descriptions of these tracks.

7. See a partial view of these tracks (figure 26) in my earlier book,' The Jurassic, A new beginning ?', 2003

8. Lockley & Hunt in 'Dinosaurs Tracks',1995, indicates similar tracks up to 7 cm in width in other regions and at approximately the same stratigraphic level - tracks still grouped today under the *Brasilichnium* classification.

9. This important matter is discussed in my earlier book along with a picture of these larger tracks (figure 36).

10. The 'Squirrel' site is covered in details in my earlier book, The Jurassic, A new beginning ?, 2003, including various photographs, graphics, and discussions.

11. Lack of association with components or organisms outside the realm of paleontology and ichnology, still today plagues the determination of the trackmakers. When zoological, geological, biological factors among others, are left out of the determination, the profile of the trackmaker is incomplete, sometime falsified. Voorhies, in 1975, was the first to point to this problem, then Hasiotis in greater details in 2002. Analytical methods being my field, I have been aware of it for several years, more so since the discovery of the Navajo mammal burrows. I will deal with this problem in a later chapter.

12. Pluvial episodes in the Navajo Sandstone were first investigated and documented by David Loope and Clinton Rowe in their paper, 'Long-Lived Pluvial Episodes during Deposition of the Navajo Sandstone', published in the Journal of Geology in 2003.

13. Prof. Lockley (along with me) investigated the 'Squirrel' tracks on March 21, 2004. He had never seen any of this type before, but speculated that a similar type may have been documented earlier in Argentina and at the same geological level. At publication time Lockley had not confirmed nor rejected his earlier speculation.

14. On my last visit to this remote site, I took away a fairly large broken piece containing several of these casts, first to study them, then transfer this rare evidence to an accredited museum. Upon Prof. Lockley's suggestion it was transferred to the Dinosaur Tracks Museum at the University of Colorado, Denver, in 2004
.

15. A photograph of that large track, along with one of the *Brasilichnium* types can be seen in my earlier book '- figure 28 & 29.

16. A photograph of this trackway (taken up the slight sloping ground, with tracks wetted for better visual acquisition) can be seen in my earlier book; figure 33.

17. Please refer to #12. These numerous wet horizons (pluvial - swamps, springs, small lakes?) in the Navajo Sandstone were first noted by Francis 'Fran' Barnes in 1997, then later by me in early 2000. Barnes later wrote about them in one of his books suggesting, then unknown *springs*, complete in one case, with numbers of petrified trees (ancient conifers). This matter was followed-up by Judith Parrish and Todd Shipman (U. of Arizona) who produced, among others writings, a paper, 'Limestones Within the Navajo Sandstone and Their Climatic Significance', published in Canyon Legacy, 2005, a publication of the Dan O'Laurie Museum of Moab. Related studies of same were made, and still are, by Len Eisenberg (an independent researcher from Oregon). Other related studies were made by Don Rasmussen, a geologist from Longmont, Colorado. As per this date, the pluvial origin of these wet horizons seems undeniable. However, their deposition, thickness, and particularly their extension, are still far from being understood.

18. From 'A Guide to the Fossil Footprints of the world', 2002, Martin Lockley, Lockley & Peterson Publication. Page 46. Also on the same page are the oldest known mouse-sized mammal tracks from the Late Triassic of Colorado.

19. Upon their discovery, these tracks were tentatively documented as *Brasilichnium*. In May 2004, during a formal investigation, Prof. Lockley chose to by-pass them, under the premise that

they were unknown in the track record. Dr. Gerard Gierlinski (Head of the Polish Geological Institute in Warsaw, Poland) and I, however concluded they were indeed of the *Brasilichnium* type but made in an extremely wet surface. Following that formal investigation I returned to the Mile 11 site (yet unseen by anyone) and uncovered identical but slightly smaller tracks. These *Brasilichnium* tracks cross a very wet surface, leaving tracks nearly identical to the ones uncovered at the Sovereign site.

For the uninitiated, tracks originally made in wet-damp sandy surfaces weather in *rounded dish-like impressions* under severe erosion. And this applies to both tridactyl (Theropods-Ornithopods) and mammal tracks. However, when preserved in wet to extremely wet surfaces, *tridactyl tracks* always retain an impression of the middle digit, albeit many a time in poor state of preservation. As uncovered at both the Sovereign and Mile 11 sites, tracks made by mammals in similar surfaces only leave a circular (rounded) impression. These impressions are not *dishes*, but flat 'rings' left by the sucking effect of the feet as the animal moved forward. These kind of 'rings' are logical since most mammals during that period had rounded-oval feet (*Brasilichnium* type).

20. At the time, the Moenave site *(Laoporus* tracks), documented by Prof. Lockley, was the only one of any importance in this region, primarily due to its location in the upper Wingate deposit of the very Early Jurassic. In the Navajo Sandstone however, the 'TOW' and 'Two-Slabs' sites were, again during that time, major discoveries of similar if not greater importance than the Moenave one.

Multitude of small Brasilichnium tracks uncovered in 2000 at the Breaking Wave site. Upper Navajo Sandstone.

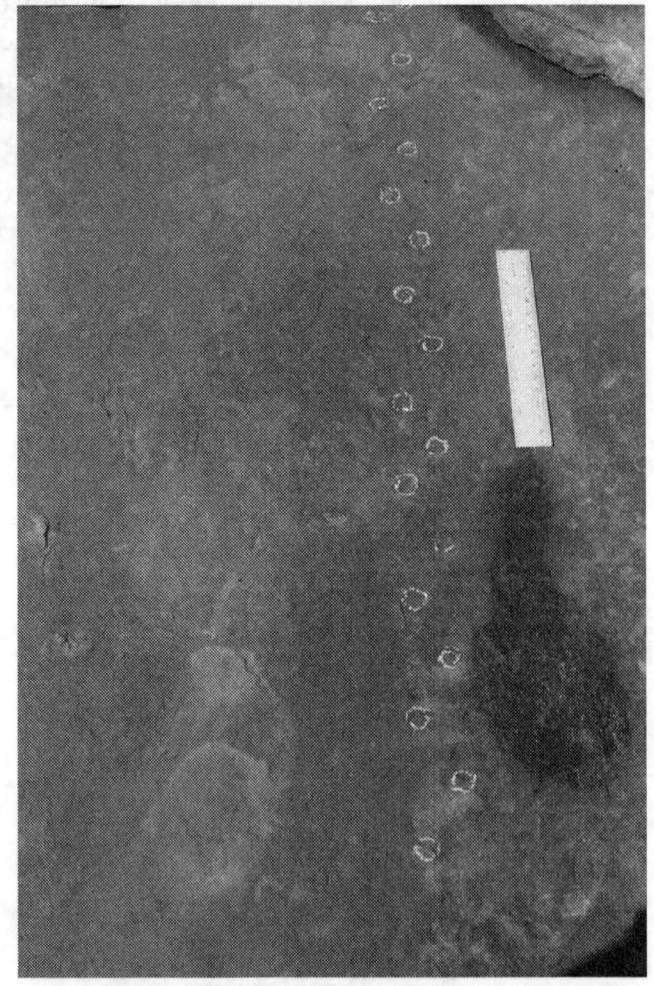

Trackway of a very small mammal similar in size and configuration to the 'Squirrel' type, but eroded. Upper Navajo Sandstone.

Trackway of a small early mammal in an upper horizon of the Kayenta Formation, middle Early Jurassic.

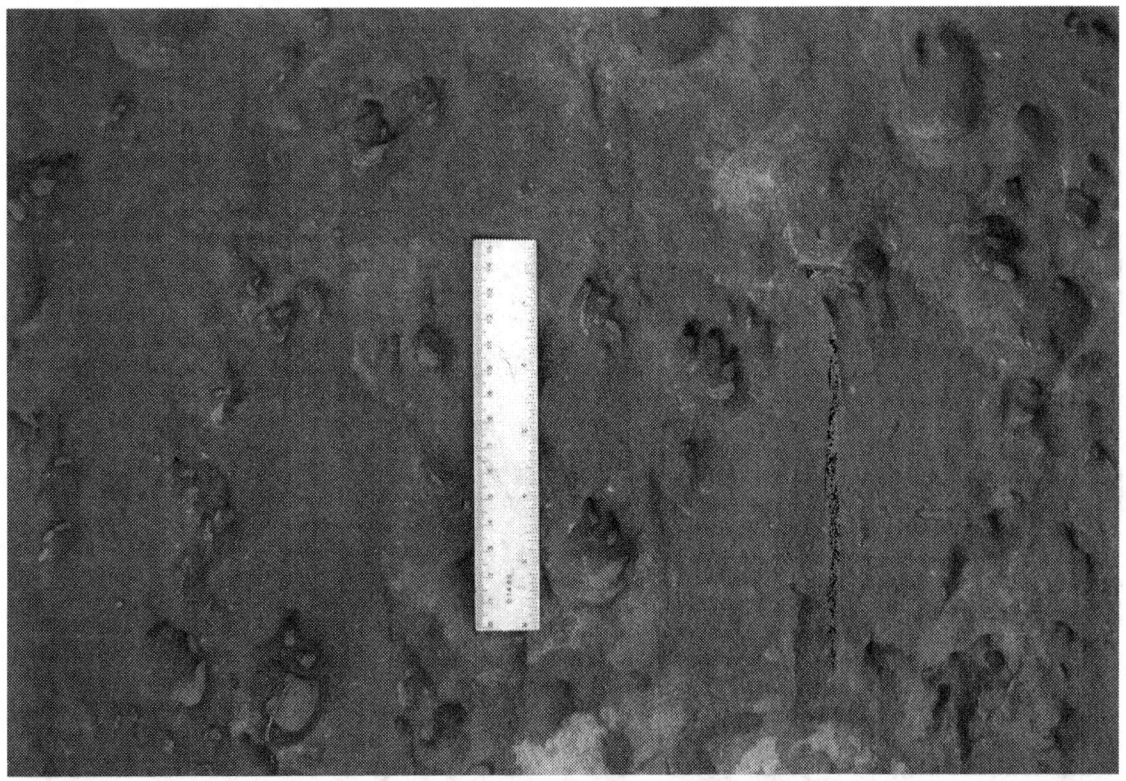

Typical and fairly common Brasilichnium tracks in the Upper Navajo Sandstone, upper Early Jurassic.

Brasilichnium trackways. At upper left center of picture (wetted area) are entrances and trampled 'highways' made by either very small early mammals or insects. Upper Navajo Sandstone.

An extremely well preserved 'Squirrel' trackway, a new type of track attributed to very small early mammal. Upper Navajo Sandstone.

'Raccoon' tracks, a new type of track tentatively attributed to some kind of early mammal. Upper Navajo Sandstone.

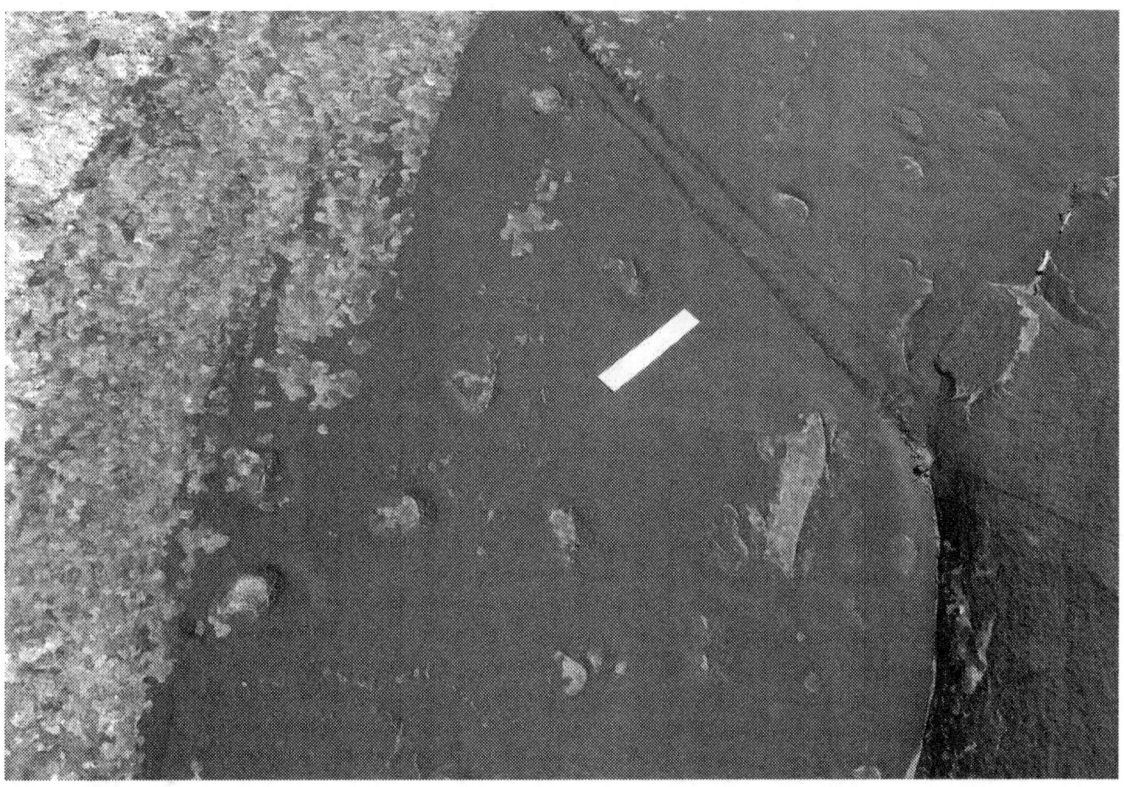

Trackway of a fairly large early mammal, and similar but smaller tracks attributed to the same species, uncovered in the Moab-Entrada Tongue of the Entrada Sandstone. Upper Middle Jurassic.

Typical Brasilichnium tracks recently uncovered in-around the very large Dewey Bridge Burrow. Upper Navajo Sandstone.

Chapter III

Mammal burrows in the Early and Middle Jurassic of Southeastern Utah

A brief account of their discovery

In 1997, Francis 'Fran' Barnes, a field researcher in geology based in Moab, Utah, came across strange cylinders, some lying around, others sticking out of a Navajo Sandstone wash. After finding similar ones elsewhere, he speculated they could be burrow-casts but when he described them to a geologist friend, they were shrugged off as 'root casts'. Nevertheless he included a picture of them in his book, "Canyon Country Prehistoric Life", published in late 1997, then again in a booklet titled, "Navajo Sandstone – A Canyon Country Enigma", published in late 1998. In both publications he mentioned them as 'burrow casts' most likely made by some kind of vertebrate.

In 1999, while studying the Top-of-the-World mammal site in the upper Navajo I also came across these strange tubes, but instead of sticking out of a ledge or a wash, they were scattered around an exposed horizon near the bottom of the small butte, where this site is located. Unlike Barnes, I thought they were anomalies or perhaps petrified roots. A while later I mentioned them in an email dealing with tracks to Laurie Bryant, the then Regional Paleontologist for the BLM in Salt Lake City. She didn't know what they were, but suggested they may be some kind of erosional features since vegetation was extremely rare in the Navajo deposit.

After teaming up with Barnes in late 1999, he mentioned to me several places containing these strange tubes. We then proceeded to investigate them together, but I was neither impressed nor interested, since my focus was solely on fossil tracks. Barnes did however report them to geologist friends, but nothing came out of it. In spite of the lack of interest, or dismissal as 'erosional features' (1), he continued to note them in his field surveys.

My real involvement only began in mid-2000. By then it had become obvious to me that Barnes' determination, as something else than 'erosional features', was indeed correct. By very early 2001, some of the puzzling aspects of these tubes were somewhat solved (mostly by Barnes), resulting in a tentative prototype for all of them. Further study however became severely hampered by the prototype's inflexibility. Nevertheless by late 2001 Barnes, his wife Terby, and I had uncovered a fairly large number of these sites, most of them of very small sizes with only a few tubes if that, to mark their presence, all located in the Navajo Sandstone in-around the Moab area. Then, due to their similarity and increasing numbers, backed by continuing lack of interest from anyone, Barnes finally put aside his search for them (2), and returned to his studies of unknown springs, also in the Navajo deposit (3). I did the same, but returned instead to my own studies of mammal tracks in the Moab region.

The discovery in early 2002 of the outstanding 'Squirrel site' in the upper Navajo changed all that. The trackmaker was clearly a burrower, almost immediately inferring that these 'tubes', at least some of them, must be of mammalian origin. I mentioned this to Barnes. He concurred with this speculation, but his priorities by that time being elsewhere, I found myself alone in the pursuit of these would-be mammal burrows (4). Not an ideal situation, but from there on, my sole focus was upon the inspection and determination of these burrows. This led to closer inspections and measurements of the tunnels' diameters, which in turn, led to studies of substrate densities, potential water sources and vegetation, tracks, location within aeolian dunes, etc.. This is when these burrows ceased to be 'one-dimensional' - that is visually similar and 'lifeless' - and began to emerge as a major component into a 'new' Navajo Sandstone, already shaken by the discovery of increasing numbers of trees, springs, and wet horizons, thanks to 'Fran' Barnes geological knowledge and field acumen.

Unfortunately my observations did not stimulate much interest among my own colleagues, and I was left on my own to substantiate such projections. By then however a string of mammal track discoveries, backed by vastly more sophisticated inspections of geologic horizons, had permitted me to uncover the second and unknown aspects of these burrows: They were no longer restricted to crumbly or not ledges, or eroded 'washes', as per the original prototype, but scattered around exposed flatish areas. In order to differentiate these two types of burrows I coined the name 'Class I' for the original and easy to spot 'ledge-wash types', and 'Class II' for these harder to detect ones. By very early 2003 the Class II types had extended the numbers and dispersion of these burrows to considerable proportions. So much so that any other origin than a mammalian one would seem unlikely. Further, invertebrates burrows studies made by specialists came to the surface, indirectly supporting a mammalian origin, at least for many of them (5).

Later on in 2003, a new kind of burrows came to the surface. Now referred to as 'Class III' types. Along with the Class I & II types, they are currently challenging the conventional view of the Jurassic of the southwestern United States and beyond. Their discovery and forms are described under the Class III paragraph.

In late 2003, early 2004, these burrows took a new turn outside conventional paleontology and ichnology. Via correspondence and discussions with biologists and zoologists friends of mine (some at the C.N.R.S in France, my country of origin), the biological, social, fossorial, etc., profile of these still unknown burrow makers began to emerge. Then, in early 2004, a British biologist, Colin Egan, recently arrived in Moab, took the biological aspect into his own hand and upon inspecting these burrows came to the same conclusion, that they could only have been made by social vertebrate species, i.e., early mammals of some kind (Colin Egan is a member of the original investigative team).

In March 2004, Prof. Martin Lockley took a brief look at one of these burrows but was unconvinced they were of mammalian origin, instead suggesting termites or crayfish burrowers. He was however unsure of his own suggestions, and subsequently asked Spencer Lucas, curator of paleontology-geology at the Museum of New Mexico, to check them out. On April 23, Debra Mickelson, a paleontologist from Denver, Colorado, and a friend of Prof. Lockley, also took a brief look at the same above burrow. A visit hampered by extremely bad weather. She speculated that the upper part was of crayfish origin, with the lower and main part most likely made by some kind of early mammals. She didn't follow the matter beyond that visit.

On April 27, Spencer Lucas, along with Tamsin McCormick (a Phd in geology), Colin Egan and I, upon inspecting a number of these burrows, concluded that they were irrevocably of mammalian origin. He returned on July 5, and upon a more thorough field investigation decided to present this discovery to the forthcoming GSA Convention. He then wrote, with the support of the investigative team, Abstract # 77249, and presented it at the GSA Annual Meeting in Denver, Colorado, on November 7, 2004. At that Meeting, I alone manned the Poster-Abstract. During that time, several geologists-paleontologists came to discuss the matter with me, most skeptical, a few supportive.

At the GSA Regional Meeting in Grand Junction, Colorado, May 23, 2005, Spencer Lucas, this time by himself, encountered skepticism and rejection of the discovery and very little support, if any. Don Rasmussen, a senior geologist from Longmont, Colorado, came to see some of these burrows on May 27, 2005, following the GSA Regional Meeting, and concluded that at least some of them must be of mammal origin. He suggested that Steve Hasiotis investigate these Navajo burrows to separate the mammalian ones from the invertebrate ones. Shortly after Steve Hasiotis accepted my invitation to do so, and eventually came to Moab for a four days investigation, November 25-28, 2005. Steve Hasiotis is an Associate Professor at the University of Kansas, Dept. of Geology, Lawrence, KS., and an expert in invertebrate and vertebrate burrows.

Meanwhile, Spencer Lucas informed our team that in spite of the skepticism and rejection he had encountered at the GSA Regional Meeting, he would complete the formal Manuscript following the original Abstract, to be presented at the GSA Annual Meeting in Salt Lake City, October 16-19, 2005. Due to a combination of a heavy work load and negative peer pressure he wasn't able to follow through by that date. However, following this GSA Annual Meeting, he reiterated his interest to produce a Manuscript in the near-by future, and an informal side-bar for this book no later than the end of November 2005. This side-bar is included in the following pages.

During the months following the GSA Regional Meeting, I made a related discovery of major interest: Sets of fairly large de-watering pipes in the Upper Navajo Sandstone, pipes unknown to most scientists. Don Rasmussen confirmed their origin but, at that stage, only the ones with large diameters (around 3 feet or over). Later, during his visit, Steve Hasiotis not only re-confirmed them as de-watering pipes but dramatically enlarged their number by uncovering smaller ones ranging in diameter from 6 inches to a foot or larger. He then proceeded to double if not triple the number of roots in-around these Navajo burrows, including very small ones of unknown origin. He also uncovered various traces of invertebrates (termites, etc.,) in-around these roots and burrows. In effect, the burrows discovery was now enlarged to include the first vegetation, and lots of it, ever found in dune-sand (as opposed to trees in-around ephemeral oasis), and unknown de-watering pipes contradicting the conventional view of the Navajo Sandstone, especially its upper part, as a dry, desolate, and vegetation-less period.

At publication time, the second but vastly more formal investigation of these burrows and related discoveries, is scheduled to take place in 2006. This investigation, or to be exact, series of investigations, is scheduled to include a much wider field of academics and professionals, most likely involving both the BLM and the State of Utah paleontologists. Also, the earlier mentioned Manuscript is now scheduled for publication by mid or late February. Should finalization be completed by that date it will be included in an Addendum to this book. If not, it will be included in a separate publication along with other pertinent manuscripts. This Manuscript, along with supportive ones, will be presented at the 2006 Geological Society of America (GSA) Annual Meeting.

Abstract # 77249
Presented at the GSA Annual Meeting
In Denver, Colorado, November 7, 2004

Georges P. Odier, Spencer G. Lucas, Tamsin McCormick and Colin Egan, Moab, Utah,
New Mexico Museum of Natural History and Plateau Restoration

Northwest of Moab, Utah, the Lower Jurassic Navajo Sandstone contains tetrapod burrows at four distinct stratigraphic levels. The most extensive burrowed horizon is a +/- 2 m thick interval 24-26 m above the base of the Navajo and can be identified over tens of square kilometers. These burrows have circular cross- sections and begin as shallow, low-angle diagonal shafts that lead to multiple terminal chambers. Branching is irregular and variable, no burrow linings or scratch marks are evident and burrow walls are mostly smooth. The burrow cross-cut bedding and have a fill that is identical to the host rock. Most burrows have a 10-20 cm diameter, and total lengths are up to 60 cm, though most are much shorter. In many places, the burrows form complex mazes that are concentrated in mounds that are 1-2 m in diameter and 0.3-0.9 thick. Locally, the upper part of the burrowed interval also contains rhizoliths, which are readily distinguished by their knobby surface texture, relatively small diameters (<10 cm), distinct (calcareous or siliceous) infilling, vertical orientation, and in some cases, downward branching. The tetrapod burrows from the Navajo Sandstone resemble therapsid burrows from the Lower Triassic of the South African Karoo. Therefore, a therapsid (probably a tritylondid) apparently produced most or all the burrows in the Navajo Sandstone. The extensive burrowed horizons in the Navajo Sandstone indicate multiple episodes of stable, relatively wet landscapes during which a profusion of small tetrapods thrived. This refutes previous ideas of a relatively barren Navajo landscape and also indicates an abundance of Early Jurassic therapsids in the American West that is not documented by body fossils. (End)

As per this date uncovering of new burrows in the Navajo Sandstone is so routine that they are almost 'collateral discoveries' to the vastly more difficult searches for mammal tracks. These burrows now extend to the upper Middle Jurassic (Moab-Entrada deposit). In the Navajo Sandstone, they currently extend 50 miles east of Moab, Utah, to Page, Arizona, 200 miles away.

From these above comments and descriptions one could easily assume that these mammal burrows are part of the current scientific data and incorporated in any types of Jurassic studies. Unfortunately, not the case. I chose the following document (among others) to illustrate the sometime stunning disparity between field and academic work, and to illustrate how burrows are still perceived in most of the scientific community. This document speaks for itself so I have included it verbatim and its entirety:
Web Page http:// www.montana.edu./news/ 102743786.html (an educational web site of the Montana State University).

Researcher investigates possible burrows at Egg Mountain.

Small mammals may have been scurrying beneath the dinosaurs that clomped around the Choteau area some 75 million years ago, says a Georgia paleontologist who thinks he's seen signs of their burrows.

Anthony Martin first noticed what could have been mammal burrows when he visited Egg Mountain two years ago. He was invited by Dave Varricchio, a Montana State University-Bozeman paleontologist at the time. Hoping to return next summer, Martin said, the discovery of tiny mammal fossils would support his theory that he really saw burrows.

"If these are indeed mammal burrows, it shows mammals had a subterranean lifestyle back in the Mesozoic Era," said Martin, a paleontologist at Emory University in Atlanta. " Even though some lines of mammals died out, some continued to use this sort of behavior."

Jack Horner, the MSU paleontologist who gained international recognition for his dinosaur discoveries at Egg Mountain, said crews have found a lot of mammal material at Egg Mountain over the years, so he wouldn't be surprised if Martin was right.

"I actually hypothesized that the mammals were living in burrows (We have a mammal in a burrow in our Egg Mountain exhibit here at the museum)," Horner said, referring to MSU's Museum of the Rockies.

"But it would be pretty cool if it could be demonstrated with fossil evidence," Horner continued. "Fossils such as this help to reconstruct the paleoecosystem, which is one of our ultimate goals in paleontological studies."

Martin presented his findings at the MSU last fall during the 61st annual meeting of the Society of Vertebrate Paleontology. He said recently that he realized the significance of his observations while going through the paleontological procedure of describing outcrops he had seen at Egg Mountain.

The outcrops, he explained, contained sand-filled tubes that were originally thought to be inorganic nodules. But the tubes were all 7 to 9 centimeters across, ran parallel to the ground and "didn't seem random at all." Three of the tubes intersected at a junction. In other words, the tubes looked a lot like burrows a modern ground squirrel might make.

"It was one of those circumstances where a scientist should always sit down and describe everything first before jumping to an interpretation, " Martin commented.

Martin hasn't examined the insides of the tubes yet. The fossilized structures are plugged with material that seems to have come from above, perhaps when a river or levee overflowed. But the fragmentary remains of mammals have been found near the tubes, Martin said. The mammals could have been marsupials. Some appear to have placentas.

If the tubes were burrows and they were homes to small mammals, it will enrich the picture of prehistoric life at Egg Mountain, the scientists agreed. Martin said researchers have justifiably focused on the dinosaur nesting sites and could have overlooked the tubes, as well as the significance of insect cocoons found in the same area.

"It makes a more complete ecological picture when you start thinking about it," Martin said of the mammal theory.

Martin said he may return to Choteau next summer while traveling West to visit new sites Varricchio scouted out in Nevada and Oregon. Martin is a senior lecturer in paleontology and ichnology in Emory's environmental studies department

Posted by Evelyn Boswell for 7/23/02

Note: Upon reading this above Web site I sent emails (followed by a letter) in early December 2003 to Anthony Martin at the Emory University in Atlanta, and similar ones to Jack Horner at the Montana State University at Bozeman informing them of the multitude of these burrows in the Early-Middle Jurassic of southeastern Utah and suggesting their possible interest in such discoveries. As per this date, I have yet to receive even an acknowledgment from any of them. One would think there would be an interest, at least from Anthony Martin the discoverer of these Cretaceous burrows, but in the real world of paleontology anything mammal is either shoved on the back burner or ignored. As per my correspondence.

We'll let the Society of Vertebrate Paleontology debate if the above Cretaceous critters were also nocturnal and insectivore, and return to the *real* mammals.

Class I burrows

The original type

Most if not all are located on ledges, whether be in small or large canyons, mesas, petrified dunes, or small rocky drop-offs. Inward from these ledges recent erosion has cleared the rocky surface of sand, dirt, and debris, so tunnels, entrances, and other evidences of these burrows can easily be seen. Vice-versa, their *inward extension* is difficult if not impossible to detect under the sand-dirt-debris cover that usually lies a short distance from the ledges.

These ledges are cross-sections of these burrows. In many places, the same burrow can be seen on both sides of small washes or canyons. But the never ending erosion has also taken a heavy toll on these cross-sections. The reason for their generally crumbly look with chunks of tunnels, stones, and various debris lying at their foot. This is where undeniable evidence usually lies. Strange-looking 'cylinders' or 'tubes' that in reality, are *casts* of the original tunnels.

When these burrows were buried under the next horizons surface water seeped through carrying with it minuscule particles of whatever materials that eventually filled these tunnels with a solid infilling. Being made of finer materials these in-fillings hardened to a much higher density than the surrounding host rock, and as erosion washes away the much softer sandstone they stick-out of these ledges or lie around broken-up. As a rule these 'cylinders' are easy to spot, not only due of their strange form, but because in many places they are of a different color than the host rock. They range from whitish-gray, reddish-orange, to a very dark green suggesting an extremely wet and marshy upper horizon. Most of them however are of a similar color than the surrounding sandstone.

When first inspected these casts were thought to be smaller versions of the original tunnels. This was based on potential shrinkage due to the drying-up of the cast, and/or the gradually heavier compression produced by following horizons (?). This speculation however was opening the door to convenient tunnel sizes that could house any known or projected early mammals. In 2002 I investigated this matter in some depth using studies of contemporary rodents affecting the strength and safety of older earth dams (6). While such comparative studies are far from definitive they did however indicate the opposite of the shrinkage theory. In older earth dams tunnels made by muskrats (or similar rodents) were, and still are, a safety issue: these tunnels in fact increase in diameter when water or seepage fill them, constantly weakening the structure of these dams. Pending further studies I believe that both observations are probably correct, thus negating each other, resulting in casts that are somewhat the same size as the original tunnel.

In many of these Class I burrows, tunnels stick out of the ledges and are easy to see from a distance, while their counterpart on the flat surfaces inward from the ledge can either be embedded, or broken and laying around. For this reason they are easy to find, and for the most, are the ones Barnes and I uncovered during the early days of our investigation. This is the type Barnes first came across back in 1997, while mine in 1999 were of a different type, the Class II, as follows .

Class II burrows

a newer but harder to locate type

A 'hole is a hole' goes the saying thus why bother with a 'Class II' type when a 'burrow is a burrow'? True enough, so why bother to go beyond the already established and relatively well documented Class I types?

Because Class I burrows are only a *partial view* of the mammals at hand. When Barnes and I first uncovered these burrows skeptics could point to the fact that they were somewhat isolated, scattered here and there, many small to very small in area, giving an impression of unimportance, in turn supporting the notion of 'rare, 'nocturnal' and 'insectivore'. By giving them names, we further emphasized this notion in the eyes of skeptics. In paleontology, particularly in ichnology, giving names to sites is very common if not standard procedure. But in these two disciplines a 'site' means a very small area, sometime down to a 2x3 foot slab. So, in the imagination of the skeptics, these 'burrow sites' were the same as in paleontology-ichnology. Very small, thus 'rare'. As we now know there is no such thing as a 'burrow site', only *burrowed horizons* within a given area.

The Class I burrows is where Barnes and I parted company. He returning to his earlier priorities, me 'holding the candle' on these barely lit burrows. The discovery of the Class II type was not the product of wandering around looking for something new, but an interpretation based on statistical and zoological projections. An abstract. But abstract speculation is one thing, field realities another. In the Navajo Sandstone in-around Moab, no one, including Francis 'Fran' Barnes who had explored that deposit for 25 years, had ever come across anything that even vaguely resembled any kind of burrows in *inter-dune areas*. Yet, according to my calculations, burrows must be present somewhere in these 'inland areas', since the Class I burrows located on ledges had to extend forward and backward beyond their narrow strips exposed by recent erosion.

My discovery, in late 2002, of the 'Mile 16 site', still today the largest Class I burrow in the Navajo Sandstone, stimulated me to thoroughly inspect not only the immediately adjacent area but its surroundings within a couple miles radius. After three days of rather harsh investigations this is what I uncovered:

1. A fairly large *playa* located on top of the small mesa dominating that site. Careful inspection of the well-preserved slabs lying around that playa did not - to my surprise - produce any tracks. However, along the slopes surrounding that playa I discovered several *trees* some of them still embedded in the playa's horizon.

2. Three (new) Class I burrows. A small one located close to the bottom of an Entrada Sandstone cliff, the other very large but stratigraphicaly much lower and located along the edge of a steep canyon. Both approximately ¾ of a mile east of the 'Mile 16 site'. Another, very small and eroded, was uncovered about 2 miles south. Then an inspection of the bottom of the various ledges and steep cliffs, part of the 'Mile 16 site', revealed burrows at various stratigraphic levels.

3. Half mile east of the site a very small, badly eroded, but fairly long ledge sticking out of sand and debris attracted my attention. It was thoroughly inspected but did not produce any evidence of tunnels or other indications of a burrow. However that ledge's general appearance did suggest some kind of badly eroded burrow (7). Speculating that it could be I began a thorough inspection of adjacent small 'patches' of slickrock recently exposed by rain water. In one of these I uncovered the badly eroded remnant of a tunnel. Stimulated by that discovery I extended my searches into several 'patches' in-around that area, and then into a slightly higher ground where the slickrock is largely clear of sand and debris. This is where I uncovered the *Class II* burrows.

What earlier I had thought to be a flattish or slightly rounded area of uninteresting slickrock, turned out to be an *eroded burrow*. Not at first-sight to be sure. But under careful inspection, here and there, still embedded or lying around in the slickrock, were petrified tunnels, similar in sizes and configuration to the hundreds present in-around the ledges of the 'Mile 16 site'.

Thus the difference between Class I and Class II burrows is not in the evidence that is essentially the same, but the *location*. That day, a new page in field research had begun. Mammal burrows were no longer confined to easily spotted ledges, they are also present in *inter-dunes areas*. Hardly a revelation since they had to be there. The reason why they were not spotted earlier is, because most of them are under sand or eroding away in exposed slickrock.

These Class II have changed the earlier 'burrow sites' into *burrowed horizons*. These burrowed horizons now expand the presence and numbers of mammals into hitherto 'devoid of life' areas of the Navajo Sandstone, and previously presumed 'deadly desert'. Their expansion is still in its infancy due to the immense territories that now must be investigated - or *re-investigated* - today still with very little or no help from anyone. However their limited expansion has already been dramatic:

For instance the 'Mile 16 site', earlier documented as nearly one mile-long but very narrow Class I type, has now expanded eastward (inland) into a field of burrows covering several acres, then across a small canyon, 3/4 mile away, where two Class I burrows are present at approximately the same stratigraphic level. Southward, the original Mile 16 site now extends 2 more miles in a continuous fashion, along the edge of the mesas, across their faces, then inland anywhere between 400 to 1,000 yards (and probably beyond). A burrow of gigantic size, and one that must also have present at the lower and upper stratigraphic levels since 4 of these burrowed horizons have already been identified.

Another example of the Class II expansion is the 'Bartlett Wash site', originally a Class I burrow located on top of petrified dunes, and no longer than 50 yards, at most. Today, it's a *burrowed horizon* that expands westward for at least 1 ½ miles, then north and south for around ¼ mile (so far). Although it is difficult, even with surveying instruments, to pin down the exact stratigraphic level of the various and separate burrows that apparently connect under large or small extends of sand, one thing is certain: There must have been hundreds if not thousands of early mammals living at that stratigraphic level, in that area alone. A breath-taking discovery.

What the Class II burrows have done, and still doing, is to inform us that these Navajo burrows where not isolated and small affairs but *inter-connected over extremely large areas*. A method of survival linked to fossorial-colonial mammals whether be in contemporary times or in Mesozoic ones. Today, more so in the future, discoveries of small or large burrows anywhere, immediately indicate a large presence of early mammals in-around the area. In other words, one burrow means many inter-connected ones although they may require a fair amount of field work to locate them (as I found out). Should the geology at hand and erosion preclude their preservation we must assume they were there since none of these early mammals could have survived in isolated locations (migrations and accidental relocations included).

Admittedly it's going to take a lot more than my limited investigations to decipher the actual or estimated extension of these already immense burrows - particularly by geologists who can pin-down various stratigraphic levels. Regardless of the outcome my own investigations are good enough to be at least in 'the ball-park'. And this already translates into *huge numbers of early mammals present in the Navajo Sandstone alone*.

Class III burrows

not quite the present but certainly the future

Note:
My use of sometime lengthy and very personal explanation to describe how these burrows were uncovered, especially the following Class III, is both necessary and informative. For field researchers unaware of these burrows it shows my past difficulties in locating and identifying them. Since they are going to be faced with the same problems these personal explanations should be of help in avoiding some of my early travails.

For the ones first coming in contact with this type of burrows we begin with an appropriate understatement: By comparison, the Class I & II are a 'piece of cake'. Class III are in conventional geologic parlance the 'Kings of Erosional Features'. And I must say, when I first looked at them, that there is a lot of truth for such dismissals. Strange areas, sometime flat, sometime 'convoluted', some with ledges, some with nothing more than sand-dirt-debris washed away from their surface by recent storms, some with some kind of solid or eroded black 'desert varnish' contrasting with the pastel hues of these desert landscapes, etc..... Strange places among the variety of phenomena that adorn southeastern Utah, and hardly a landscape that excites the imagination. But most important of all: *Not easy to spot evidence that, once upon a time, they were burrows inhabited by multitudes of mammals.*

This affair began in the vast expanses of the Moab-Entrada deposit (upper Middle Jurassic) north-east of Moab. Gray, dreary, and in Summer deadly expanses of slickrock in which my intuition told me there must be hiding some kind of 'Mother Lode' - mammal tracks. A correct intuition as it turned out. However, back in early 2000, I had come across what then appeared to be remnants of very small petrified trees (or brushes?). With the discovery of Class I burrows in the Navajo Sandstone these 'petrified trees' were re-classified as petrified *tunnels* - the first indication that mammals had indeed been present in that forlorn deposit. (and the reason for my mammal tracks investigations).

Then aware that some kind of burrows may be present somewhere in this deposit, I began to pay more attention to the shape of the dreary slickrock, including depth and direction of the 'ravines' cut through it by ancient or recent water erosion. This is how I discovered that the northern segments of the Moab-Entrada deposit had ravines and small, but fairly deep canyons running somewhat diagonally across the 6 % grade, while the southern segments had much smaller ravines running straight down the grade. That observation led me to take a much closer look at the mounds lying about or lining the sides of the 'ravines' in the southern segments.

The only thing I could come-up with is that many of these 'mounds' had a strange 'convoluted' surface. Probably some kind of 'erosional feature' dear to geologists. However being allergic to off-the-cuff 'erosional features' (and for good reasons), my curiosity kept me hammering at a better explanation. I then speculated that these mounds must be of higher density than the conventional marine-deposited sand of the Moab-Entrada (left by the retreating Middle Jurassic Seaway). This speculation was, if not 'dead-on-the-mark' at least close to it : To my surprise I discovered that the upper sections of the Moab-Entrada (eastern boundary) were *not* marine-deposited but *cross-bedded aeolian dunes*. A different ball-game.

This is where matters stayed until the discovery of the Class II burrows in the Navajo Sandstone. By that time mammal burrows were solidly in place, with 'erosional features' flying out of the window. And with it came a new perception of these strange Moab-Entrada mounds. Perception yes, but unless backed by solid evidence these 'mounds' would remain 'erosional features'.

But solid evidence only existed in the Navajo Class I & II burrows. So any evidence supporting a new kind of burrow would have to come from *related factors*. Remembering the small and badly eroded 'ledge' in the vicinity of the 'Mile 16 site', I had earlier investigated with no evidence that it may have been some kind of badly eroded burrow (although its overall appearance did suggest such possibility). I returned to that site and spent almost a full day inspecting every inch of it, forward and backward. Again no tunnels or indications of same. However the color, the visual density, and the generally rounded remnants of that crumbly ledge clearly indicated some kind of moist horizon, perhaps the remains of a spring ? Since the 'Mile 16 site' was only half-mile away I decided to make a comparison between the two to try to elucidate this moist horizon 'thing'.

Inspecting some of the most eroded and crumbly aspects of the Mile 16 site, by walking along the foot of the small cliff, I uncovered similar horizons complete with rounded remnants and similar colors: Given its location and similarity with this undeniable Class I type, I concluded that the tunnel-less ledge must have been a burrow - a badly eroded one with no tunnel evidence but with *characteristics* similar to Class I and II. This is how this new type of burrow came to the surface, and to identify it from the others I named it *Class III*.

Now knowing their rough configuration I immediately began a search for them in other areas of the Navajo Sandstone, and sure enough I uncovered a handful in a very short time. However, one of these was going to be the key to unravel the then would-be but later *major* Moab-Entrada Class III ones. Therefore a description of this investigation is in order.

It is now referred to as the 'SF 300 site' (meaning the Sand Flats BLM Area east of Moab, and its approx. 300 yards length). When I first looked at it (through binoculars) all I could see was a crumbly ledge above a very small canyon, with broken patches of desert varnish on its upper surface extending inward to an adjacent close-by ledge. Nothing to get excited about since such places are quite common in Canyon Country.

On a second trip I noticed more patches of desert varnish at the same level as the 'crumbly ledge', but this time extending a fair distance down the small canyon. I decided to investigate, and after nearly 4 hours of crawling around this dreary place I concluded that this vast area must have been a burrow - but with no solid evidence to back it up.

This large site was a challenge, and with my professionalism on the line, I had to 'crack it'. So instead of wandering again around that site looking for evidence that did not exist, I went back to my earlier discoveries of Class I & II burrows looking for clues, a time-consuming process that finally paid-off. Then I returned to the SF-300 site, but this time armed with photographs, measurements, and various details of these undeniable burrows. That time, no doubt, this site was a burrow, but a highly eroded one beyond 'conventional' interpretation. Still no signs of tunnels to back it up. I was looking at a large burrow nobody would 'buy', my pride bruised, along with the distressing sound of geologists howling 'erosional features'!.

I went back a fourth time, this time to take 'technical' photographs for later reference. Looking for the 'perfect angle' to capture the essence of that site, I hiked along the crumbly ledge up to the place where it merges with a somewhat vertical and less eroded slope. This is where I found the solid evidence: On the now better preserved ledge were a few petrified tunnels. I then hiked down the small canyon, and in its bottom I uncovered a few more, some still embedded in the host rock. The 'SF 300 site' was indeed a burrow. A Class III.

Note: in 2005, I uncovered a large Class II type approx. 30 feet above the 'SF-300'.

Armed with this knowledge, I returned to my 'mounds' in the Moab-Entrada. Anticipating another series of intellectually-demanding and physically dreary explorations I chose first to get an overview, a 'feel' for this immense field of slickrock lying in the southern segment. I hiked toward a high point half-mile away up the slickrock (the Moab-Entrada has a 6% tilt). This high point is on a narrow 'tongue' of the upper Summerville deposit carved by recent erosion.

The top of this 'tongue' like many others in this deposit is 30 foot or so above the slickrock, its bottom washed away by rain water. Near the high point, and looking down at that sharply defined interface, I noticed a fairly large greenish-brown 'patch' contrasting vividly with the gray-white of the slickrock. Although I had seen a few of these 'patches' before, most of them small and visually of little interest I felt the time had arrived to check what they could possibly represent (8). A wise move as luck (God?), this time was on my side, something that rarely happens in field investigations. Upon reaching that strange 'patch' I was almost immediately 'in the money'.

Numbers of very small but inter-connected tunnels still embedded in the surface, lying among scattered remnants of vegetation. Next to them, a small Class III burrow, then again, scattered tunnels sticking out of the surroundings ravines. This was irrefutable evidence that mammals were indeed present in that supposedly lifeless deposit.

The diameters of the very small inter-connected tunnels range in size from 3 cm to 4 cm. Most likely made by some kind of invertebrates although this interpretation is still tentative at this date. The rest are small tunnels ranging in size from 5 to 7 cm, and identical to similar ones documented in the Navajo, clearly mammal-made.

The petrified vegetation is still undocumented at this date. Some is well preserved and resemble some kind of bush while others appear to be remnants of roots commonly found in today' swampy areas or at the edge of rivers and stream (9).

Now knowing that these greenish-brown 'patches' were remnants of some kind of wet or swampy grounds hospitable to vegetation, insects, and mammals I began a systematic exploration of the lower part of that southern segment where most of these patches are located. Days of investigation had mixed results: Most of these patches had nothing to offer, many very small and badly eroded. However, I uncovered three somewhat similar to the original - one very large in size, approx. 25 yards square. The other two much smaller but containing scattered vegetation (bush remnants similar to the ones at the first site) and what appeared to be some eroded remnants of surface tunnels (?). The very large one however contains both very small surface tunnels (insects?) and vegetation, and on its periphery I uncovered a few mammal-made burrows identical to the ones in the first patch. At that site I also made a very interesting discovery: On a small and exposed cross-bedding near mammal-made tunnels are several well-imprinted *bird-like tracks* (10).

As educational as these above investigations were, they did not elucidate the 'mounds' mystery, nor they contributed anything beyond swamps, vegetation, insects, and burrows suggesting a scattered perhaps rare mammal population. They managed however to keep my interest alive in these vast expanses of dreary slickrock. I was about to 'throw-in the towel' when on October 18, 2003, at 4:36 pm (MST), I discovered the 'smoking gun'.

By major discovery standards, it was not a spectacular one, far from it. Walking back to my vehicle, parked on the Willow Spring Trail 200 yards from a remote entrance to Arches National Park, I decided to take a 'last look' before going home, at one more of these hundreds of strange 'mounds', this one only a few feet from the trail. Sticking out of it were *several petrified tunnels*. These strange mounds were not 'erosional features' but *Class III burrows* ! Amen.

These strange 'convoluted' mounds are only a form of *bioturbation*, a geological term for horizons reworked by various organisms. In that case, these horizons were reworked by mammal-made burrows, the reason for their roundish aspect, emphasized by erosion. Also the reason why they were preserved, albeit in very poor condition. Made in moist soil, their reworked density was slightly higher that the surrounding host rock, thus better preserved.

It was 'down-hill' from there on. No more 'dreariness' or 'hallucinations' in the hot sun. With these petrified tunnels now imprinted in my brain I began to 'see' them, here and there, most of them barely visible. Worse, in most of these mounds none could be found. But this time around it was a different ball-game. Tunnels or no tunnels these bio-disturbed mounds now indicated either burrows or burrow-disturbed areas. And to make sure they did I went around many of these mounds searching for tunnels...and to date, every time I made such searches I uncovered at least one of them. Later on, I became so confident that I actually 'challenged myself' in finding them in badly eroded surfaces, and so far always have (but too time-consuming to prove a 'point'). Until rain and snow ended all explorations here is a summary of what I have uncovered so far:

1. In the southern segment 8 cross-sections were made from top to bottom (11). In each and all these cross-sections petrified tunnels were uncovered in enough of these mounds (even in some badly eroded areas) to strongly suggest that this vast area was either a burrow field or a combination of burrows and foraging areas.

2. In the middle segment of the deposit roughly 3 miles long and somewhat centered in-around the Dalton Well 4x4 trail only 3 cross-sections were made (one partial). This segment is generally badly eroded with only indications of scattered burrows. No tunnels have been found to date. However, it is in this area that several trackways made by fairly large mammal-like animals were uncovered (The Sovereign Site, as described earlier). The same area where a substantial number of bird-like tracks were also uncovered along with scattered remnants of vegetation.

3. In the northern segment, the last segment (12), somewhat centered around the Klondike Bluffs Trail, only two cross-sections were made, one extensive. No burrows of any kind were uncovered in this wide expanse of extremely eroded slickrock (approx. 3 miles wide). However in its upper and less eroded northeastern corner indications of burrows are present (including a handful of petrified tunnels, remnants of vegetation, and bird-like tracks). Also, two separate and very small sites containing eroded but still visible tracks of a fairly large mammal-like animal (identical in configuration to the Sovereign site type) were uncovered in-round the Klondike Bluffs Trail area. Lastly, 'patches' of what may be highly eroded remnants of burrows or mammal-disturbed areas are scattered here and there on that segment. Many of these patches are clearly bio-disturbed and of a different color than the surrounding slickrock (some 'yellowish' and fairly easy to spot).

4. Another but separate segment of the Moab-Entrada was briefly investigated. This very large segment (approx. 2 x 1/2 mile) is located west of Hwy 191 and directly above the Hidden Canyon area. This separate segment is highly eroded and somewhat similar to the middle one described earlier. A very narrow cross-section was run on its northern side, but no farther than ½ mile. Indications of badly eroded burrows (or burrow-disturbed areas) are scattered here and there but none convincing enough to be solidly interpreted as mammal-made. However, in the middle of the cross-section, I uncovered several petrified tunnels sticking out a small and eroded ledge confirming some sort of mammal presence.

What constitutes a Class III burrow ?

Anything that cannot be explained via conventional methods of investigation or lack of proper explanation in geological literature. Thus, to understand Class III burrows you must first understand the Class I & II types. In summary, the Class III types are only badly eroded versions of the Class II. This is where knowledge of local and regional geology comes into the picture; that is the effect of ancient and recent erosion. While erosion in general terms can easily be seen in Canyon Country, its impact on burrows and similar 'proxies' to the presence of ancient life is barely known today. For instance, I uncovered that most if not all traces of burrows disappear in steep or straight-down cliffs or mesas, thus nearly impossible to detect (if there). This observation, repeated on several occasions, is based on solid evidence. In many a place a Class II disappears inside such cliffs or mesas then re-appears at exactly the same level on the other side. Another example is the unexplained presence of flat or flattish slickrock, lying many a time between cross-bedded dunes, or small buttes or mesas. Unexplained because these flatish, strange, and uninteresting areas are not explained in conventional geology. Many a time these strange areas represent the *foundation* of a burrow that was totally wiped-out by erosion. Or nearly wiped-out, because in many of them some remnants of tunnels and tunnel entrances can be found in less eroded 'niches'.

A major component permitting detection of these Class III burrows is a recent discovery. Vertical tunnel entrances, some of large diameter and sticking out of the sand cover. Numbers of them have since been found in both Class II and Class III burrows. In most cases erosion has rounded them into strange cones suggesting some kind of termites or similar insect construction. They are not, only entrances to lower and horizontal burrows of mammal origin. These strange-looking rounded cones are many a time all is left of obliterated burrows in-around flattish slickrock. In some burrows vertical entrances are still preserved in their original state, vertical and rounded casts, some with tunnels still attached to their lower part. In most cases however, severe erosion had reduced them to conical mounds. Better anchored in the slickrock than horizontal tunnels-casts, many of these entrances-cones were able to survive, albeit in reduced form.

Another facet of the Class III types is the 'crumbly' but 'roundish' look of some ledges, as uncovered at the 'SF-300' site. This aspect is a true 'erosional feature'. Severe erosion has wiped-out the tunnels (casts) then 'rounded' the host rock into a bio-disturbed appearance (as per the 'mounds' of the Moab-Entrada). In dry aeolian horizons, like the Navajo or the Entrada Sandstone, cliffs and ledges crack and fall down vertically, with sharp edged slabs and debris lying at their bottom. When moist, as per many of these burrowed horizons, they hardened at a higher density and were shaped into 'roundish' forms by ancient and recent erosion. When such 'roundish' ledges are uncovered, the best place to look for evidence is inland from the edge, where still embedded tunnels and entrances had a better chance to survive severe erosion.

Overall, Class III burrows are still a long way to be understood, but their emerging presence is further enlarging the already immense size of the Class II types. No doubt, they are going to be the key factor in uncovering burrows in other regions of the Pangaea Continent. For this reason, fellow scientists should help unveil their mysteries following familiarization with the Class I & II types. Exploration for any of these burrows is no longer a hit and miss proposition but a planned discovery.

Why it took so long to detect these burrows?

Since I am in a better position than anyone to answer that question here are some answers. The main one rests in the self-inflicted limitations of paleontology and ichnology, and by inference, geology. Field researchers who, like me, depend on these sciences for guidance and studies, were informed in no uncertain terms that any kind of vertebrate life, with the exception of a few Theropods and possibly some rare to very rare primitive mammals, did not exist in these dreary sand seas. No vegetation either, except perhaps around some vague oasis. Thus, anything beyond the above was either some kind of 'erosional features' or the product of the imagination.

Francis 'Fran' Barnes and I were not the first to see these burrows. They had been around for hundreds of years but nobody knew, nor were interested to find out what they could be. During the 1940's and 60's Uranium Boom in-around Moab, Utah, some, perhaps many geologists had in fact came across them. At two old drill sites many burrow-casts are still lying around their periphery, and at one large burrow site (the Paris site), drillers used several of these burrow-casts to make fire rings. Paleontologists, geologists, amateur 'trackers', etc., who earlier or later must have come across at least some of these casts never 'saw' them either. However, when they were missed by such authorities as Spencer Lucas, Steve Hasiotis, or Martin Lockley, you know you've got a problem.

This problem is the same as it was for me earlier on. Even by late 2000, when by that time I had seen dozens of these burrow sites, my reaction was still, 'So what?'. My interest was strictly focused on tracks, not vague 'tubes', or 'erosional features' that abound in the Navajo Sandstone. If this wasn't enough to forget about them, lack of interest among paleontologists-geologists was the 'coup de grace'. Field researches, especially the difficult ones, are largely stimulated by the desire to uncover items of interest to the scientific community-at-large. When this interest is lacking, very few if any researchers are going to spend their precious time in the field looking for or study items of interest to no one, I concluded.

These 'tubes' did not come 'alive', until I made the connection between them and mammals via the discovery of several mammal track sites. When this connection was established, at least in my mind, then I began to 'see' these 'tubes' everywhere. Today, with these burrows and burrowed horizons well implanted in my brain, I am capable to detect them as far as 100 yards, and twice that distance with binoculars. 10 feet to 200 yards is the difference between 'tubes' of interest to no one, and mammal burrows today challenging the conventional Jurassic. When one knows that difference you'll be amazed how quickly you'll be able to 'see' these burrows in the field.

Notes on modern mammalian burrowers

With *no mammal burrows documented in the entire Mesozoic Period,* the only reference at our disposal are the Lower Triassic ones uncovered in the South African Karoo, made and inhabited by Therapsids, and the Cenozoic ones after the KT Extinction. They are even extremely rare in that Cenozoic Period, only surfacing in its later part, the Miocene and Pliocene.

However, this information turned out to be incomplete. Upon meeting with Steve Hasiotis in late November 2005, he informed me of fairly similar burrows he and his colleagues had documented in the Late Triassic and Late Jurassic, and subsequently sent me two documents pertinent to their discovery and investigation. These documents are included in the References section. The title of the main one is, "Vertebrate Burrows from Triassic and Jurassic Continental Deposits of North America and Antarctica: Their Paleoenvironmental and Paleoecological Significance" (Ichnos, 2004). In this document, Steve Hasiotis, Robert Wellner, Anthony Martin and Timothy Demko, interpret these burrows as made by fossorial mammals, vertebrates with burrowing behavior similar to mammals and mammaloids. Their extension however, is minimal by comparison to the gigantic ones uncovered in the Navajo Sandstone. Nevertheless, their documentation by experts in 2004, strongly supports our original team's contention that the Navajo burrows were also made by similar fossorial mammals.

In addition to the above and new documentation, modern mammalian burrowers can also be used as a starting point, and working 'backward' from there. The following is a condensed version of M. R. Voorhies work on the subject , a truly amazing piece of research published in 1975 under 'Vertebrate Burrows', by Springer-Verlag New-York-Berlin itself under the title of 'The Study of Trace Fossils'.

In this below version, any modern vertebrates burrowing in wet-moist horizons or near-by water (alluvial, palustrine, etc.) are left out of the list, since the Navajo and Moab-Entrada burrows are *only found in dry aeolian horizons*, any moisture found or perceived in them being of *pluvial* origin. For the record these Classes are, Amphibia, Reptilia (all except lizards), and Aves (most birds; some do burrow in dry horizons but the matter is academic since birds did not exist in the Early Jurassic). The only Class left is Mammalia.

Families of Mammals known to include a least one species that digs its own burrow:

Order Monotremata
Spiny anteaters & platypus.

Order Marsupialia
Oppossums, marsupial mice, marsupial moles, bandicoots, wombats, kangaroos

Order Insectivora
Tenrecs, shrew-otters, golden moles, elephant shrews

Order Edentata
Armadillos

Order Pholidota
Pangolins

Order Tubulidentata
Aardvarks

Order Lagomorpha
Rabbits, pikas

Order Carnivora
Dogs, bears, raccoons, weasels, civets, hyaenas

Order Hyracoidea
Conies

Order Artiodactyla
Pigs, peccaries

Order Rodentia
Sewellels, squirrels, gophers, pocket mice, beavers, springhaas, New World mice, mole rats, bamboo rats, Old World mice, dormice, dzalmans, jumping mice, jerboas, Old World porcupines, cavies, agoutis, chinchillas, hutias, coypus, hedge rats, tuco-tucos, chinchilla rats, cane rats, African mole rats, and marmots.

The Order Rodentia was kept last on purpose. From this list one can see at once that the *rodents* are by far the most numerous and prolific burrowers of the entire Mammalian Class. And these are the same or similar rat-like animals we know from body, cranial, jaw, and teeth fossils from the Jurassic and the Cretaceous. On that list also, one can see similar burrowers, like rabbits, golden moles, marsupial mice, etc., that millions years ago could not be separated from the general rat-like appearance of early mammals.

From this contemporary burrowers' list, we can draw a determination that is certain to raise a lot of controversy, a determination first invoked by Voorhies, then later by me (13) : *Contemporary rodents are living fossils*, at least the smaller prey species. Their lineage in both the Navajo and Moab-Entrada deposits clearly rests in this multitude of smaller burrows. A fossorial lifestyle that could not have changed from the Early Jurassic to contemporary times, since, as small prey species, it was then and still today their main, perhaps only method of survival.

This above aspect suggests that the main mammal lineage rests inside these Navajo burrows. Larger species, known from their tracks and body fossils (like *Kayentatherium*) appear to be a lesser advanced form of early mammals. In turn, they appear to be evolutionary dead-ends, whether they disappeared sometime during the upper part of the Jurassic, or, as some claim, at the end of the Cretaceous. If so, it would indicate that the *two branches* of the main mammal lineage lived side-by-side inside these Navajo burrows, or near-by in somewhat similar ones: The 'prey branch', as per small contemporary rodents, and the 'evolving branch' that later evolved into a variety of mammals. When this differentiation took place between the two branches is unknown but it did take place. However, the recent discovery of *Repenomamus giganticus* and *robustus*, both large carnivorous mammals, coupled with the tracks of a large mammal-like animal in the Moab-Entrada, point to a branching as early as the Late Jurassic, and no later than the Early Cretaceous. This branching may have been minimal at that time, perhaps only affecting carnivorous species, that must have been present in these burrows (14). But since it *did* take place, we can at least assume that some of the larger burrowers were also on their way as a separate branch, albeit in minimal ways.

Moving to the K-T Mass Extinction, there is no doubt whatsoever that any kind of fossorial species had the best chance to survive that extinction, and in fact, did. This leaves larger mammals like *Repenomamus* 'out in the cold' so to speak. Extinct, unless able to dig their own burrows. Did they ?. Similar contemporary animals do, at least enough for temporary shelters and the raising of their young. We won't know until we uncover larger burrows or shelters. Whatever the case may be, we can at least assume that the *main branch* of the Mammal Class (evolving and prey species) must have been *fossorial* at least to a degree up to the K-T Extinction. That's what these Navajo and Moab-Entrada burrows are pointing out today.

THERAPSID BURROWS IN THE JURASSIC NAVAJO SANDSTONE

Spencer G. Lucas

In late April 2004, I met Georges Odier in Moab, Utah, to examine some of the vertebrate footprints he had discovered in the Triassic-Jurassic Wingate Sandstone. At the end of my brief visit, Georges showed me an extremely odd outcrop in the Lower Jurassic Navajo Sandstone, northwest of Moab. Intercalated with typical crossbedded sandstones of the Navajo is an approximately 2-meter thick interval honeycombed with what Georges believed to be burrows, probably made by mammals. These structures were new to my experience, and I knew little about vertebrate burrows at the time, so I could not fully evaluate Georges' assertion as to the origins of the structures.

The bizarre structures in the Navajo Sandstone piqued my interest, so I reviewed the (limited) technical literature on vertebrate burrows (see especially Groenewald et al., 2001). Then, in July 2004, I returned to Moab for a more extensive look at the "burrows." That work convinced me of the validity of Georges' conclusion that the structures he found are vertebrate burrows. Thus, they have subcircular cross sections and begin as shallow, low-angle diagonal shafts that lead to multiple terminal chambers. Branching is irregular and variable, no burrow linings or scratch marks are evident, and burrow walls are mostly smooth. The burrows cross-cut bedding and have a fill that is identical to the host rock, but the filled burrows preferentially weather from the host rock. Most of the burrows are 10 to 20 cm in diameter, and total lengths are up to 60 cm, though most are much shorter. Most significant are the "nests" that some of these burrows form. These are complex mazes of burrows concentrated in mounds that are 1 to 2 meters in diameter and up to 0.9 meters tall. Horizontal burrows connect some of these mounds to each other. Indeed, the pattern of the mazes and burrows is strikingly similar to the nests and burrows of East African mole rats (see Jarvis and Sale, 1971).

If these are not burrows, then the most likely alternative is that they are rhizoliths (root casts). However, significantly, the burrowed interval also includes rhizoliths, which can be readily distinguished from the burrows. These rhizoliths have a knobby surface texture, small diameter (less than 10 cm), distinct (calcareous or siliceous) infilling, vertical orientation and, in some cases, downward branching.

Who (what kind of animal) made the burrows in the Navajo Sandstone near Moab? No skeleton of a burrow maker has yet been fished out of one of the burrows, so identification of the burrow maker(s) must be by inference. The Navajo burrows resemble some of the therapsid burrows described in the literature, and they also resemble the burrows of some living mammals, as noted above. This suggests to me that therapsids probably made the burrows in the Navajo Sandstone. My reasoning is influenced by our knowledge of Early Jurassic mammals and therapsids. Those mammals (for example, *Sinoconodon*) were too small to have made many of the Navajo burrows, though a mammal origin for some of the smaller burrows cannot be totally ruled out. Early Jurassic therapsids are mostly tritylodontids, a group of rodent-like animals whose range of body size easily accommodates the Navajo burrow sizes. Tritylodontids are extremely abundant in some Early Jurassic fossil assemblages, such as that from the Lufeng Formation of Yunnan, China. And, a tritylodontid skeleton has been reported from the Navajo Sandstone. Therefore, it seems a reasonable inference that a large number of tritylodontids made those burrows in the Navajo Sandstone near Moab.

The extensive, burrowed horizon in the Navajo Sandstone discovered by Georges Odier opens up a new window on an Early Jurassic landscape. Instead of a vast wasteland, the Navajo sand sea was home to a substantial biota. Tritylodontids were herbivores, so many tritylodontid burrowers implies sufficient vegetation for food. Furthermore, tritylodontids were not rare during the time of Navajo deposition, but instead quite abundant. The Navajo dune sands preserved few body fossils, but if we look at the Navajo trace fossil record a true picture of life in that long lost desert will emerge.

References

Groenewald, G. H., Welman, J. and MacEachern, J. A. 2001. Vertebrate burrow complexes from the Early Triassic Cynognathus zone (Driekoppen Formation, Beuafort Group) of the Karoo basin, South Africa. Palaios, 16: 148-160.

Jarvis, J. U. M.a nd Sale, J. B. 1971. Burrowin and burrow patterns of east African mole-rats Tachyoryctes, Heliophobius and Heterocephalus. Journal of zoology London, 163: 451-479.

Spencer Lucas Phd,
Curator of Paleontology and Geology
New Mexico Museum of Natural History & Science
Albuquerque, NM.

Analysis

The past and still going-on dismissal of these burrows by conventional academics or professionals as mammal or mammaloid made, is first anchored in a **bias** against mammals, and second, in the rejection of, or lack of knowledge in related disciplines, the primary one being Ichnology. Bias or lack of interest in early mammals is so rampant that it is not worth a discussion. But lack of acquaintance with **Ichnology** is inexcusable, either demonstrating a limited scientific knowledge, or rejection of that discipline as a scientific component. The up-to-date fossil tracks record of the Navajo Sand-Sea only shows the presence of two types of vertebrates: dinosaurs and mammaloids. Tracks attributed to invertebrates or other kind of vertebrates are rare to extremely rare, some next to non-existent. Therefore, these gigantic and complex Navajo burrows could only have been made by very small to extremely small dinosaurs, or mammaloids.

In the reptilian class, this record further shows a Navajo Sand-Sea almost totally dominated by **tridactyl** tracks attributed to either Theropods or Ornithopods. To date, no small to tiny tridactyl tracks, that could be attributed to **burrowing reptiles**, have been uncovered in that deposit. This taking into account that the search for mammaloid and small to tiny tridactyl tracks is identical in terms of difficulty and competence. Further, and assuming that such small to tiny tridactyl tracks had been missed by dedicated and competent field researchers like myself and others, it would assume that small to tiny **bipedal** Theropods or Ornithopods, most likely infants, some the size of mouse, were capable of digging these gigantic and complex burrows. An assumption that stretches the imagination. First, there were no reasons for infant dinosaurs to dig such tunnels for their protection, in the animal kingdom whatever protection is always provided by the parents at least initially, and second, bipedal animals cannot dig burrows with their feet, they have use their beak or their teeth. Some modern birds do burrow with their beak, but there are no evidence whatsoever in the ancient and modern world that similar bipedal animals could or can burrow with their teeth. In the modern animal kingdom only certain types of **moles** can burrow using their teeth. Until someone can provide evidence that tiny Theropods or Ornithopods, could dig such gigantic burrows with their teeth, the only animals that could have dug and lived in these burrows were **early mammals**. They were the only ones with the anatomy and the social-biological complexity needed to do so. And this since the Early Triassic, as confirmed by relatively recent but largely unknown discoveries of their burrows – with their numerous skeletons still buried in them.

As for the proposition that semi-aquatic invertebrates like crayfish, lungfish, toads, salamanders, etc., could have burrowed and lived in such arid conditions without the watery environment necessary to keep them from almost immediate desiccation and death is preposterous at best. None of these burrows have ever been found in such arid and water-less conditions in ancient and modern times, and none will ever be found. These semi-aquatic species are 'living fossils' with a record stretching back 250 million years.

As for termites or similar invertebrates, their burrows are actually present inside or around these mammal burrows, as it should be, since they were surviving on the same vegetation as the mammaloids. But their tunnels, mounds, and radiation, have already been documented, albeit in recent years and unknown to most scientists. For anyone to assume or ascertain that these gigantic and complex Navajo burrows were made by termites or similar invertebrates is to show a lack of knowledge in Ichnology and worse, a lack of knowledge or lack of interest in scientific documentation made by peers in similar or related disciplines. Termites were indeed present as it should be but their tunnels, etc., are not similar nor even close to the burrows made by early mammals. These Navajo burrows were **unknown** to the scientific world until now. Their sole similarity is to Early Triassic burrows already documented as mammaloid-made. To dismiss them or lump them as termites-made is not a demonstration of scientific knowledge, but a lack of it.

At this early stage, these numerous and gigantic burrows must be seen from a higher perspective than arcane debates over the types of early mammals who made and inhabited them, a matter that will be taken care of when body fossils are uncovered among them. From this higher perspective we can draw a general conclusion of the highest implications: If these multitudes of early mammals could live and thrive in the most arid and desolate region of the entire Pangaea Continent, they must have been present by the millions, in the temperate and hospitable ones. From this hard to refute projection, *early mammals were the dominant species in term of numbers, right at the beginning of the Early Jurassic*, and *not* after the K-T Extinction as presumed.

From this now known and/or projected multitude of early mammals present at the end of the Early Jurassic, to assume or assert that their numbers were reduced to rare levels at the end of the Cretaceous Period, including down-sizing to mice-like animals, is in great need of explanation. Since carnivorous dinosaurs were unable to keep their numbers down by the end of the Early Jurassic, they could not have done so later in the Jurassic and the Cretaceous, periods, vastly more hospitable to mammals than the beginning of the Jurassic. A better explanation, and most likely the correct one, is that early mammals were just as numerous and diverse, perhaps even more so, at the end of the Cretaceous as they were at the beginning of the Jurassic. The difference between 'rare' and 'numerous' is on one side solely based on a fossil record that is proven to be inadequate for such task, and on the other, sophisticated projections anchored in a number of abstract and physical proxies. Until now, the fossil record has been a reflection of the dominance of dinosaur studies in paleontological and geological circles. With the renewal of interest in early mammals raised by this and related discoveries, a shift toward our Jurassic ancestors should up-grade the fossil record beyond lackluster field performances when it comes to Mesozoic mammal species. A shift already noticeable in China, the country that in recent years has provided the international scientific community with an array of new mammal species that, combined with these Navajo burrows, today offer a different and more accurate vision of our ancestors.

Via their numbers these burrows point to an *equivalent amount of vegetation*. From various remnants (roots/rhizoliths) this vegetation appears to be some kind of 'surface type'. Whatever it was it must have been quite prolific to sustain such numbers of animals year-around. Further, the size of these burrowers (based on tunnel diameters) raises serious questions as to what they could actually reach or were able to cut down to feed on - since grass is still unknown in the Early-Middle Jurassic (15).

None of these tunnels could have been made, more so preserved, in loose, dry sand. They were made in *compacted horizons*. And in these sand-seas, compaction could only have been made by water, *rain* or high humidity. This compaction-rain factor explains the presence of vegetation, which in turn explains the presence of this multitude of early mammals. This vegetation, today only known via its numerous roots (rhizoliths) uncovered inside or on the edges of these burrows, was not an *ephemeral one* but an everlasting type necessary to sustain this multitude of animals year-around. And to be plentiful and everlasting this type of vegetation needed an equivalent amount of rain water, not short, periodic types confined to ephemeral plants, but pluvial episodes scattered throughout the entire year. This vegetation-pluvial episodes factor in turn explains the *compaction* of the horizons in which these animals built their city-like burrows. It also explains the presence and preservation of flatish slickrock commonly found in the Navajo Sandstone, slickrock in which numbers of burrows are present. This flatish slickrock was either a horizon in which these animals burrowed, or the *foundation* on which these burrows were built. Either way, it was compacted by rain water that later lithified at a higher density than the following dry or dryer deposits, the reason for their preservation. This water-induced lithification was the same for the burrow-casts as it was for the slickrock. The only difference being that the in-fillings of the burrow-casts, made of seeping fine materials, eventually lithified at a higher density than the much larger sand grains of the slickrock horizon. This difference can easily be seen in Class III burrows where eroded or not burrow-casts are still embedded in the faster eroding surface of this slickrock.

The largest tunnel diameters so far uncovered indicate a vast majority of small burrowers, right down to mouse-like animals. The largest somewhat similar to contemporary badgers.

At this stage it's impossible to tell if these multitudes of burrows were inhabited by an equivalent numbers of burrowers. Some may have been vacated for various reasons, the most common being depletion of foraging areas and search for new ones, albeit close by due to the size and vulnerability of these animals. Estimated population densities are usually based on the quality of the habitat. Thus, using the *conventional habitat* of the Navajo and Moab-Entrada deposits these mammals should *not* be present, if so, only in minimal numbers. But these earlier and conventional habitats - especially the Navajo Sandstone - are today in serious need of revision, following various discoveries and documentations made in recent years, and continuing. These burrows only add a new dimension to such badly needed revisions. For the first time in the history of the Mesozoic, geology and its related disciplines must now adapt to, or take into consideration, the presence of animals, and not the other way around as in the past. This up-grading to inter-disciplinary levels should be of benefit to everyone, paleontology and ichnology in particular.

I predict that similar burrows will be uncovered outside the Early and Middle Jurassic of Utah, particularly in the Cretaceous, and *anywhere around the world* (16).

Summary

These Navajo burrows indicate that early mammals in one form or another were numerically the dominant terrestrial vertebrate species at the very beginning of the Jurassic, and not after the 65 mya K-T Extinction as presumed. To believe or assert otherwise is going to need explanations beyond the one-dimensional methodology of today's insecure fossil record.

The long-held theory that the rise of the dinosaurs was the sole or main factor in keeping mammals from 'exploding' prior to the K-T Boundary has been falsified by the discovery of these immense burrows. Mammal numbers, varieties, particularly their size, were not affected, or less affected, by the presence of predatory dinosaurs but by other factors such as lower oxygen levels, epidemics, etc.

The main branch of the mammal lineage must have been fossorial at least up to the end of the Early Jurassic in the southeastern-western region of Utah.

In these Navajo burrows the two segments of the mammal lineage, the evolutionary segment and the prey species, co-habited or lived close to each other, inferring that some evolutionary species must have been predatory.

The small prey species that inhabited the Navajo Sandstone and the Upper Middle Jurassic (Moab-Entrada deposit) most likely never abandoned their fossorial lifestyle, remaining barely evolved prey species from the Early Jurassic to modern times.

Due to harsh conditions and limited survival means only a very small percentage of the total Early Jurassic mammal population could have made and inhabited these Navajo burrows. The rest, the overwhelming majority, must have lived and thrived in the more hospitable regions of the Pangaea Continent.

Small prey species, better adapted to arid conditions than larger ones, made it through the even more arid Entrada deposit, re-emerging at the end of the Middle Jurassic in almost identical burrows than the ones they left behind in the Navajo. This re-emergence in almost identical desert conditions suggests that similar prey species living in hospitable regions must also have been fossorial, and by inference, that the evolutionary segment must have been also, at least partially.

Fairly large mammal species known today via their fossils and tracks may not have been able to survive the K-T Mass Extinction unless fossorial or partly fossorial. Until equivalent-sized burrows or shelters are uncovered, particularly at or near the K-T Boundary, this appears to be the case. Unless falsified by discoveries of larger skeletons and tracks immediately beyond the K-T Boundary it would indicate that the main mammal lineage was indeed fossorial, at least to a point, from the beginning of the Jurassic to the K-T Extinction, whether they lived in arid or hospitable regions of the Pangaea Continent.

Sole dependence on the body fossil record to project mammal numbers and varieties has proved to be inadequate for that task. Use of complementary 'proxies', physical and abstract, are now a must to verify or sustain complex projections.

Traces of life in the Desert...

In the area of Moab, Utah, Georges Odier has trekked across deposits of ancient sand seas of the Lower Jurassic Glen Canyon Group that in size rival vastness of the modern Sahara Desert. His goal was to assist scientists in documenting the different kinds and stratigraphic positions of vertebrate tracks and trackways mostly in the Lower Jurassic Wingate Sandstone. His journeys took him into the rocks known as the Lower Jurassic Navajo Sandstone in the upper part of the Glen Canyon Group to look for traces of life in this ancient desert. Because Georges was not academically trained in geology, he was not privy to previous studies that reported no signs of life in this ancient desert. Although the explorations of his friend Fran Barnes had found deposits what are now thought to have been ponds or natural springs. No such other deposits had been found in much of the Navajo Sandstone.

Georges, working on the power of natural curiosity and pure observation, noticed some very interesting shapes and forms in these ancient deposits that did not fit the interpretations told to him by the professional geologists and academics working in the Moab area. An acquaintance of Georges suggested he contact me because of my specialty in working with organism behavior resulting in traces, trails, burrows, nests, rooting patterns, and biolaminations—trace fossils. On his invitation and description of the traces of life he thought he was seeing in these rocks, I had to join him in the field to see for myself. Over my three day visit with Georges, I saw things that I would never had expected seeing in the Navajo Sandstone, or any other Lower Jurassic eolian deposits. These are share here with you.

Observation 1—burrow-like structures. Georges showed me an incredible number of beds that he traced across the countryside for me. He thinks these are mammal burrows because of their shape, size, and overall architecture. To me at first glance, they even looked like they could be mammal burrows! At this point, who made the burrows is not important – Therapsids, early mammals, lizards, or some other vertebrate—what is important is that he is on to something here that has been overlooked by many others. These animals, regardless of their status or classification, were present in the Navajo Sandstone by the thousands and represents evidence that the interpretation of the kind of desert the Navajo was needs to be seriously re-evaluated. Even more interesting with Georges interpretation was that he was unaware of the discovery of the fossil of a burrowing mammal in the Upper Jurassic Morrison deposit in Grand Junction, about 100 miles from Moab and 30 million years younger. Conventional thought until this point was that burrowing mammals were not present prior to the Cenozoic—the Morrison fossil says otherwise. These possible Navajo burrows need scientific explanation, for they may be as telling as the Morrison mammal fossil. Now is the time to study these structures and look for skeletons to determine if they are mammals, Therapsids, reptiles, or something different.

Observation 2—remains of vegetation. Georges took me to places where he and others found what looks like casts and partially permineralized remains of vegetation, which to me was amazing because I had heard stories of this but had never seen it. These structures and fairly large root casts were also found with the burrow-like structures. Together we spotted more of these remains as well as what appear to be root casts of several varieties, including small and thin branching kinds. Finding these structures in any other continental deposits would be lead to the interpretation of a well vegetated surface of a landscape. If these are the trace fossils of plant roots, they do not seem restricted to 'oasis' and ephemeral lakes found by Fran Barnes. Is this prolific but low-level vegetation living year-around in sand-dune with no apparent water source? Georges' discovery of these traces, along with the burrow-like structures, contradicts the conventional view of the life, conditions, and climate of the Navajo Sand-Sea, including the one I have learned through colleagues and the peer-reviewed literature. This is one more reason to professional studies of these and other deposits in the Navajo Sandstone need to be undertaken.

Observation 3—dewatering pipes. Lastly, Georges showed me some incredible giant, vertical cylindrical structures that he and Don Rasmussen found, which in other rock units have been interpreted as dewatering pipes. As we search the outcrop together we could see all sizes of these cylindrical structures that preserved contorted, disturbed, and reworked cones of Navajo eolian sediment. Some were at least 2 m in diameter; others were as small as 10 cm in diameter. These pipes seemed to have been formed during Navajo time, and were so prolific that they would be called a 'de-watering field'. Is there a relationship between these cylindrical structures, the vegetation remains, and the burrow-like structures? Is this the water source for the Navajo lakes and oasis, as well as for the traces of vegetation and the burrow-like structures? What is the source of all this water and why did it manifest itself in late in Navajo time from? These discoveries by Georges contradict the conventional view of the upper part of the Navajo Sandstone as a very dry period of this sand sea. In fact, it appears to be just the opposite. My head still hurts thinking about this and the other evidence Georges shared with me during my three day visit to Moab. Why? If these are dewatering pipes, then it means water, and plenty of it. The water and what ever the source of it may have been enough to support the organisms that produced those other traces discovered by Georges. It would also explain the presence of dinosaurs in the Navajo Sandstone based on trackways that Georges also found, albeit in smaller numbers.

It seems that Georges has found a variety of proxies as trace fossils that may be telling us the Navajo Sandstone has much more to it than we have previously thought. The burrow-like structures, root casts and steinkerns of tree trunks, and cylindrical structures are proxies for the possible existence of some sort of burrowing mammaloids or reptilians, patchy to abundant vegetative ground cover, and abundant water. Could this be true? There is only one way to find out and it will take more than three days, however. Georges Odier has stumbled across an abundance of evidence that warrants an in-depth look at the Navajo Sandstone and its interpretation as a very dry and desolate sand sea.

Stephen T. Hasiotis Phd,
Associate Professor, Dept. of Geology
University of Kansas

Chapter Index

1. 'Erosional feature' is an 'in-house' term commonly used by geologists (but also by many related scientists) to describe forms, shapes, or contents - usually small ones - unknown or poorly documented in the geologic record. Lately however this term has acquired a less than sterling reputation. It has been used, and is still being used in some quarters, to dismiss out-of-hand discoveries that are not in line with conventional theories. In scientific terms there is no such thing as an 'erosional feature' unless accompanied by plausible explanations. When unaccompanied by any explanations, or at least some reservations as to what it may be, it either shows a lack of professionalism or worse, contempt for the work of fellow researchers.

2. Francis 'Fran' Barnes passed away in October 2003 the victim of skin cancer he had unfortunately contacted during his many years as a field researcher in harsh 'Canyon Country'. However, some of his research is still available at his Moab residence thanks to 'Terby', his widow and former field assistant.

3. Francis Barnes' little known but outstanding contribution to science. Although the author of many books and reports on the subject his discoveries and implications are condensed in a 300 pages 'Journal of the Navajo', a document he never had a chance to publish due to his sudden departure. As a field note his burrows' discoveries first occurred while searching for trees and springs during wide-ranging and extended explorations of the Navajo Sandstone.

4. Francis Barnes and I teamed-up for nearly 4 years, but in different fields. He in geology and related sciences, I solely in fossil footprints, first general, then specialized in early mammals. His interest in ichnology was peripheral, usually limited to pictures and measurements, and when possible noting their location. Upon his discoveries of springs and large trees in the Navajo deposit he lost most of his interest in these then barely known burrows, compounded by the lack of interest from the scientific community.

5. Among others, Steve Hasiotis' superbly documented article published in 2002 in Palaeo, Elsevier Sciences. This article titled, 'Complex ichnofossils of solitary and social organisms: understanding their evolution and roles in terrestrial paleoecosytems', is a must read for anyone interested in the study of these mammal burrows. In essence, Hasiotis had brushed aside any possibility that these Navajo burrows could have been made by anything but small vertebrates of some kind. In the *interior* of the Navajo Sand-Sea the fossil and track record do **not** support any kind of small reptilians except infant Theropods (or Ornithopods) or small species of bird-like dinosaurs. Thus, vertebrate burrow-makers were either small to very small dinosaurs, or early mammals, whose small size and classification are already established. While mammaloids and early mammals' burrowing abilities are also fairly well established, the burrowing abilities of bipedal dinosaurs, large or small, have yet to be proven. For a discussion of the matter please refer to the Deliberations Chapter.

6. These studies were made by the Water Rights Branch of the British Columbia Provincial Government (Canada) during the 1950's in order to assess the safety of hundreds of small to fairly large earth dams built throughout the Province at the turn of the century.

7. This is the ledge that later became the first evidence of Class III burrows.

8. The reader may ask, correctly, how could I had possibly 'brushed aside' these 'greenish patches' during my earlier explorations of that deposit ? Here is the answer, and one that should be of interest to fellow field researchers: Our eyes may see 'things' but unless these 'things' - unknown tracks or unknown 'features' - are *imprinted in our brain by prior knowledge* chances are that you won't 'see' anything, your brain rejecting what you 'see' as unimportant to what you are doing. This phenomenon is known to psychologists and other brain specialists but very few persons are aware of it. One day after failing to 'see' blatant and already documented tracks on a desert varnish-coated slab I asked a psychologist (a family-member to avoid embarrassment), what this was all about. So don't feel that you may be the sole 'victim' of that phenomenon. I was too, and on a number of occasions.

In fact I had 'seen' these 'greenish patches' during earlier explorations since I had to cross at least two of them to get to two very important and then very rare bird-like track sites. But I only saw what I was predisposed to see. My brain had obviously 'rejected' them as unimportant 'erosional features'.

As to why I 'saw' these mysterious 'mounds' and not these 'greenish patches' there is a (possible) explanation: Due to my extensive field studies of *prior* Class I and Class II burrows I now believe that their configuration had 'alerted' my brain that they could also be a form of burrows. An information probably stored in the remote parts of my brain. When, by sheer curiosity (or perhaps boredom with these vast expanses of similar slickrock (?), I took the time to look at these strange 'mounds' my brain then 'regurgitated' these prior studies or visual configurations (?) unconsciously informing me that these mounds were more than run-of-the-mill 'erosional features'.

9. The discovery of this vegetation leads to an anecdote worth mentioning. I took a geologist friend of mine (whose name I won't divulge by courtesy toward his extremely conventional views of that deposit) to that site for him to inspect 'in situ' this new aspect of the Moab-Entrada. He acknowledged the burrow-tunnels but dismissed the vegetation out-of-hand. When I asked him to offer an explanation for these brush-like remains he first told me that there is no vegetation record in the Moab-Entrada geology. Unsatisfied with such a dogmatic answer I asked him what could have possibly made some of these thin branches lying about the site? '*Invertebrates*', was the answer. Then pointing out to the limbs gradually getting smaller to a point (as all limbs do) I asked him what kind of worms could have possibly made such 'tunnels' ? Caught off-balance he hesitated for a moment then came-up with one of the finest scientific explanation I have ever heard: 'Who knows? We know next to nothing about these Jurassic invertebrates and for that matter they could have come from another planet '(sic). This is how far some of these 'scientists' will go to deny any evidence that could up-set the conventional view of the subject at hand.

10. This discovery is of importance. The only ones known until then in the Moab-Entrada were a handful, most of them eroded, discovered by Francis 'Fran' Barnes in-around 1997. This new site was documented by Prof. Lockley in May 2004.

11. Two of these cross-sections were made 'in tandem' with two visiting and very helpful scientists friends of mine, Gunther Permoeller and his wife Elise from Germany.

12. Actually the last easily accessible segment that is continuous with the two others. The Moab-Entrada deposit extends further north, there consisting of two smaller and separate areas hard to access.

13. When I first proposed this possibility and projection to several US paleontologists, particularly specialists like Richard 'Rich' Cifelli (Oklahoma Museum of Natural History), it was met with disbelief if not derision. When I suggested they extend their studies to contemporary rodents, it was either ignored or rejected as a discipline (zoology) outside their own.

14. See the 'Predation' sub-chapter in the Deliberations part of this book

15. A serious question because it deals with their ability or not, to climb trees or high brush, inferring a potential arboreal lifestyle that would contradict these burrows as mammal-made. *Eomaia scansoria* from the Early Cretaceous is the first known mammal with tree climbing ability, an interpretation based on its feet structure. Whether it was arboreal or fossorial is unknown, but as a rodent-like animal it was most likely fossorial with tree-climbing ability, like many similar contemporary species.

16. And this has already begun. In early 2004 I uncovered one Class II and two Class III in the Navajo Sandstone of the San Rafael Reef, 150 miles from Moab, and this during a relatively brief exploration. Then in late 2004, I uncovered similar burrows on the way to Page, Arizona, 200 miles from Moab, also in the Navajo deposit. With a now established range of +/- 250 miles, these burrows must also be present in all Navajo-like deposits in the Western US. In July 2005, Nathan Wilkens, Phd student at Arizona State University, in an article titled 'Paleoecology of the Navajo Sandstone Interdune Deposits' (co-authored by Prof. Jack Farmer, same university) reported, along with a photograph, mammal burrows but in Arizona and under the caption, 'Burrows or tree roots?'. This article was published in the Summer 2005 issue of Canyon Legacy, a publication of the Dan O'Laurie Museum, Moab, Utah.

From left, Colin Egan, Tamsin McCormick and Spencer Lucas during the second inspection of some Navajo burrows, July 6, 2004.

Burrow-casts sticking out of a steep ledge.

Large burrow-casts eroding out of the edge of a vertical cliff.

Typical burrow-casts lying around or embedded in Class II & III burrows.

Burrow-casts and eroding burrow entrances. Notice entrances conical form.

Typical but partial view of a burrow located along the edge of a bench.

Burrow extending from foreground all the way around the mesa in background.

Burrow extending along the edge of a mesa for nearly one mile.

An immense burrow located 30 miles east of Moab, Utah.

Don Rasmussen inspecting roots of low-level vegetation located in-around the gigantic Mile 16 burrow. May 28, 2005.

Typical and very common root found in-around burrows.

Close-up of a root showing insect marks; probably termites or similar insects.

Multitude of roots still embedded inside a burrow.

Steve Hasiotis photographing insect traces inside a burrow; probably termites or similar insects. May 26, 2005.

Typical insect tunnel, probably made by termites or similar insects, commonly found in-around burrows. Notice small diameter of tunnel by comparison to mammal burrow-cast lying left of it. Also notice the almost straight alignment of the tunnel typical of derived termites (sand termites).

Insect of very small mammal burrows? This type of burrow can be found inside or outside mammal burrows, the later in cross-bedded dunes as per this picture.

Partial view of a very large Class II – III burrow located on a rocky bench cleared of debris and sand by recent rains.

From left, Gerard Gierlinski, Martin Lockley and Laura Mitchell making casts of bird-like tracks uncovered in the vicinity of mammal burrows in the Moab-Entrada deposit – Upper Middle Jurassic. March 20, 2004.

From left, Gerard Gierlinski, Martin Lockley, Laura Mitchell and Dona Gierlinski measuring a Sauropod or Ornithischian trackway in the Moab-Entrada deposit. This trackway was uncovered in 1997 by Francis 'Fran' Barnes and is the first confirmed presence of Sauropods (or ?) in that deposit. March 20, 2004.

Large burrow-casts emerging out of sand.

A well-preserved burrow-cast still embedded in the host rock.

A Class II burrow extending over 1/2 mile.

Steve Hasiotis photographing burrow-casts lying around the very large '101' burrow. November 27, 2005

Spencer Lucas photographing vertical burrow entrances at the 'Mile 16' burrow. July 5, 2004.

Large vertical burrow entrance with trunks of 5 sub-horizontal tunnels still attached to it (2 are on the other side of the entrance).

Vertical burrow entrance or de-watering pipe? Albeit located inside a prolific burrow this vertical-cylindrical structure is most likely a smaller type of de-watering pipe recently uncovered in the Navajo Sandstone.

Eroded and conical burrow entrances emerging out of the sand. These easy-to-spot whitish 'cones' are usually the first indication of an eroded Class III burrow.

A badly eroded Class III burrow with small entrances still visible on its surface. This particular burrow must have been burrowed and inhabited by very small mammals, as indicated by the entrances and the diameter of the burrow-casts still embedded in its lower horizo⁻

Effect of erosion: A broken down burrow-cast lying in the sand of a Class III burrow.

A lonely but still embedded burrow-cast indicating a burrowed horizon nearly wiped-out by recent erosion. At this location a few entrances 'cones' were uncovered farther down the rocky wash.

Effect of erosion: Barely visible but still embedded burrow-casts cross-cutting the bedding of a Class III burrow.

A badly eroded Class III burrow lying in front of the Organ Butte in Arches National Park.

A fairly preserved Class III burrow in Arches National Park with 'Balanced Rock' in the background.

Tamsin McCormick standing below a Class III burrow with broken tunnels lying at its base. Notice the light color and crumbly aspect of the upper burrowed horizon indicating a wet to very wet period, as opposed to the lower and darker one (actually dark-red) indicating a dry period.

Very large burrow-casts still embedded in a Class III burrow.

A large and branching burrow-cast still embedded in a Class III burrow.

Partial view of a burrowed horizon emerging out of the sand.

Partial view of a Class II burrow recently uncovered by torrential rains.

Colin Egan inspecting multitude of Triassic crayfish burrows uncovered in the Moab, Utah, region.

Crayfish or lungfish burrows? Uncovered by the author in 2005 in the Lower Navajo Sandstone 130 miles west of Moab, Utah. Pending confirmation they may be the first crayfish-lungfish burrows uncovered anywhere in the Jurassic.

Close-up of Triassic crayfish burrows. Crayfish and lungfish burrows are vertical, mammal ones are horizontal or sub-horizontal.

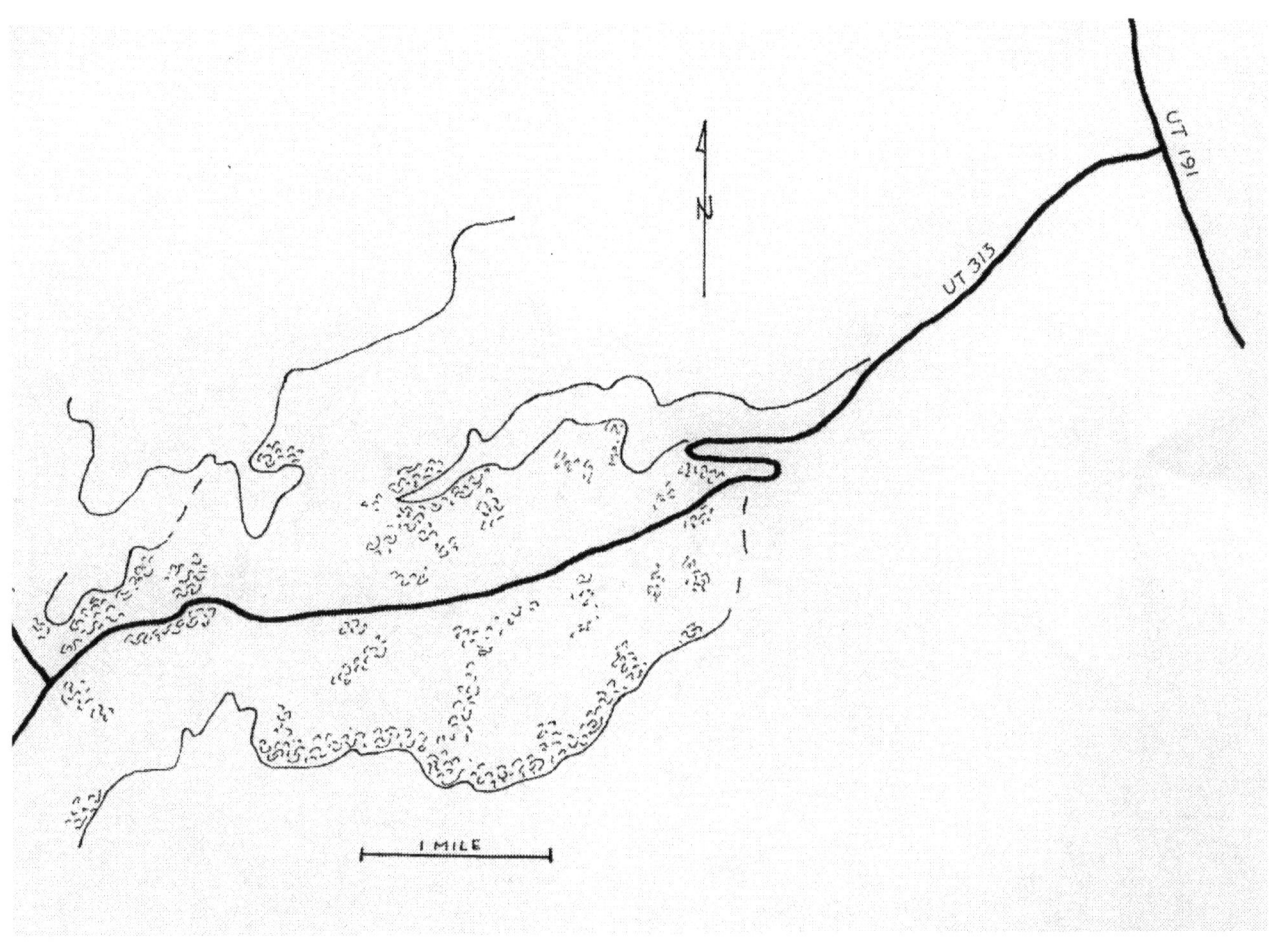

Schematic map showing the extension of the Mile 16 Burrow. Only documented burrows are shown. Recent investigations indicate that many if not most of them must have been considerably larger and inter-connected. Their numbers alongside the edge of steep cliffs is solely the product of recent erosion. Inland, many of these burrows are partially buried under sand and debris.

Chapter IV

Related Discoveries

Vegetation in the interior of the Jurassic Sand-Sea

Vegetation in the Navajo Sandstone was practically unknown until 1999. That is, in the interior part, the overwhelming part of this Early Jurassic sand-sea. Elsewhere, on its periphery, limited vegetation had already been discovered or speculated. The lack of fossil vegetation evidence was one of the main factors behind the the prevailing view of that deposit, as an inhospitable and nearly lifeless period. Oases, suggested by scattered playas (ephemeral lakes), were thought to be the only places where some kind of vegetation could have grown and survived in such an inhospitable, arid environment.

In that year, Francis 'Fran' Barnes, a field researcher from Moab, Utah, uncovered chunks of petrified trees in-around the shores of a large playa. That discovery was followed by another, reinforcing the implication of the first: several petrified trees lying around the edge of a spring, some still embedded in it. Then again followed by more trees, all in-around playas. At this time, these discoveries, trees, playas and springs, challenged the conventional view of the Navajo Sandstone, but there was resistance from conventional geologists, paleontologists, and paleobotanists. The trees were immediately dismissed, in the same manner as the mammal burrows were later. The first dismissal verged on the ludicrous. The following is a description, to demonstrate the extent to which some conventional academics will go to dismiss anything that does not confirm the prevailing view.

After several attempts to interest various scientists in these newly uncovered Navajo trees, one - a retired paleobotanist from the University of Arizona, whose name I shall not divulge, out of courtesy to his age and retirement – finally came to Moab to look at these 'so-called trees'. Upon looking at several of them, some two feet in diameter, lying around a spring, he informed the stupefied Barnes and his associates that they were not trees. Asked why, he replied: "because they are no trees in the Navajo". End of the Navajo trees, at least for a couple months. Barnes then informed this 'scientist' that in fact they were trees – conifers – as confirmed via an independent lab testing. This fellow, to his merit, returned to Moab and the site, but insisted that these now-trees could not have grown in the Navajo Sandstone, and provided Barnes and his associates (including me) with the following explanation:

"They must have floated down a river from higher grounds somewhere in the East". When asked for evidence of such a 'river' in the geologic record in front of him and the trees, he answered, "I am not a geologist". When informed by Barnes that such evidence does not exist in the Navajo, at least in this region and location, he then came up with one of the finest scientific explanations that Barnes and I have ever heard: "They must have been carried by a nasty storm and fallen from the sky".

This above illustrates the 'lack of vegetation in the Navajo Sandstone' view, right up to 1999. Fortunately for science, nobody could get sidestep these trees, now growing in number, and by 2001 they were finally accepted, albeit grudgingly, by most scientists. However, all of them were uncovered in-around playas and springs, confirming the earlier speculation of vegetated oases, and no trees elsewhere in dry or dryer dune fields. In 2003, I uncovered numbers of what appeared to be roots (rhizoliths) in-around mammal burrows, the second evidence of vegetation in the Navajo Sandstone. These roots were first confirmed by Tamsin McCormick, in 2004, using field testing techniques, and again by Spencer Lucas, then by others in 2005. These roots are of still unknown brush-like plants, that must have been extremely prolific year-round, to sustain these multitudes of early mammals. More interesting, it was the first evidence of vegetation growing and surviving year-around in, believed to be dry, aeolian dune fields. With this discovery the conventional view of a 'desolate' and 'barren' Navajo sand-sea went out of the window. Low-level vegetation, and plenty of it, existed at all levels, and in vast expanses of that Jurassic desert. Nowadays, these vegetation roots have become so common that they are only of 'collateral interest'.

To the above we add a new discovery I made in November 2005. In a fairly large expanse of exposed slickrock located on top of a butte (Upper Navajo) I uncovered the remains of a wooded area. Fragments of trees, some with trunks still embedded in the slickrock, ranging in size from 2 to 14 inches in diameter, but all located in dune-sand. The first evidence that trees were also present outside playas, springs, and wet horizons, and could grow to, at least such diameters, in dune fields – similar to junipers (a member of the conifer family) that dot the contemporary Navajo.

The above dune-sand horizon is now documented but not the trees themselves, although they are, or appear to be conifers, similar to the ones in-around playas, etc. A following and better investigation of that area by Tamsin McCormick (Phd in Geology), Colin Egan and I has located a wet horizon immediately above the sand-dune one, indicating that these trees were indeed growing in a dune field. It's too early to tell if these trees (conifers?) were growing elsewhere in the Navajo Sandstone under similar conditions.

However, we can safely speculate that they must have been. Today, that the Navajo was replete with burrows, brush or low-level vegetation, trees, wet horizons, springs, playas, de-watering pipes, etc., and further discoveries of 'desert trees', should not come as a surprise. The ancient Navajo was not a barren lifeless affair as previously thought but a 'living desert' somewhat similar to the contemporary one.

In December 2005, and again in January 2006, I discovered trees, many of large size, growing inside two playas, and not around their periphery as previously thought. Their location implying that these trees were partially or fully grown before the creation of the playa. In other words, their origin and roots were in the last dune-sand horizon, not in the limestone of the following playa. At time of publication that discovery has yet to be confirmed by geologists, but it appears to be on solid ground, as these trees are too large to have grown solely from the thin limestone deposit. In one place the trunk of a very large tree had 'busted' through the limestone layer, then petrified later in its original position.

If proved correct, this recent discovery would indicate that trees, some of large size, were indeed growing in dune-sand. Not anywhere, but in moist-humid locations. In-around playas this proposition seems logical.   For these ephemeral lakes to form the bedding first had to be sealed. This sealing process could only have made by a combination of rain water and dust-like material, somewhat similar to clay. Ergo, the flat or flattish horizons on which all these playas rest must have been moist to wet to begin with. In the Navajo sand-sea, they made almost perfect places for trees to grow, and apparently they did. If trees could grow in flat-flattish but moist-wet locations they must have been able to grow in similar places where the water-dust sealing process wasn't tight enough, or thick enough, for a playa to build on.

So far the only proof of 'desert trees' is the discovery of that wooded area in November 2005. Could field researchers elsewhere have come across similar 'desert trees'? I don't know of any, but such a possibility exists. Here however we run into a problem of field identification. Tree fossils, particularly chunks and debris of these, are difficult to very difficult to notice in sandy conditions and among other natural debris like chert and limestone, more so if the person is inexperienced in the configuration and colors of these petrified vegetative remains. Further, since these 'desert trees' did not exist in the geological record, no one in particular was looking for them, myself included until recently. I predict that many more will be found after field researchers become aware of their potential presence, especially in horizons offering signs of ancient moisture.

The last but not least question is what type of vegetation this multitude of early mammals were feeding upon? Not trees obviously, but some kind of plentiful low-level vegetation today solely represented by numbers of roots uncovered in-around these burrows. But these roots in turn seem to represent some kind of fairly sturdy brush that may or may not be what these animals were feeding on, at least as their primary forage. Their teeth configuration and small body size suggest that they instead may have been feeding – grazing being a better word – on some kind of plant-grass.

This speculation is not a loose one. A recent discovery, published in Science magazine (November 2005) has debunked the theory that grasses did not emerge until long after the dinosaurs died off. The earliest grass fossils ever found were about 55 million years old – 10 million years after the K-T Mass Extinction. Varieties of grass have now been found in the coprolites of plant-eating dinosaurs dating between 65 and 71 million years ago – the upper part of the Cretaceous. This discovery, that came as a surprise to conventional scientific circles, was made by Caroline Stromberg of the Swedish Museum of Natural History, and Vandana Prasad of India's Birbal Sahni Institute of Paleobotany.

While it came as a surprise to conventional circles it did not to the various scientists involved in the documentation of these gigantic Navajo burrows and the multitude of early mammals who inhabited them. Grass does not leave behind fossilized remains, at best extremely rare and difficult to detect imprints on wetted surfaces. In the Late Cretaceous it took 'dung' (coprolites) to uncover it. Dung, like these immense burrows, is a 'proxy' for lack of solid evidence. Under the conventional 'one-dimensional' methodology grass could not exist, and didn't. It does now, thanks to the 'dung proxy'. Whether or not grass or similar plants existed in the Jurassic has now shifted from negative assertions solely based on the fossil record to more scientific propositions based on inter-disciplinary studies.

Were some kind of grasses present in the Navajo Sand-Sea? Until further botanical studies of these roots inform us of the plants-brushes they actually represent, we cannot be sure if they were the sole foraging opportunity for this multitude of mammals. However, lack or paucity of roots in some of these immense burrows points to an 'alternate vegetation'. And this 'alternate vegetation', without roots nor fossils, could very well be some kind of primitive and unknown grass.

During his four days visit in November 2005, Steve Hasiotis almost immediately uncovered numbers of small to tiny roots within these Navajo burrows and in-around humid to wet horizons. These small to tiny roots had escaped earlier investigations and at once have enlarged the number (and perhaps the variety) of vegetative remains from already substantial to prolific. Further, he identified vast numbers of horizontal or sub-horizontal roots among burrows, some with apparent or potential termites' marks (or similar insects). These types of roots (rhizoliths) are similar in diameter to many of these burrows and in places can be visually interpreted as mammal-made. Closer inspection however, especially their infilling, quickly reveals their vegetative origin.

As per this date, Steve Hasiotis has substantiated the original interpretation made by Tamsin McCormick and Spencer Lucas, later by Don Rasmussen, that unknown but numerous brush-like plants were present in dune-sand, a remarkable discovery. It now appears that a variety of plants may also have existed beside the larger brush-like roots. It also appears that this still unknown vegetation was complimentary to the presence of large to immense mammal burrows, inferring that it only grew in humid to wet areas of the Navajo sand-sea. If so, the origin of these humid to wet sand-dune areas needs to be explained. Rainfall at first glance appears to be the only plausible origin. But rainfall does not discriminate and during that time must have been homogeneous over vast areas of that desert. The recent discovery of various-sized de-watering pipes, but more important 'de-watering fields', adds a new dimension to the origin of this water. Until geologists better investigate and decipher the complexity of these inter-connected discoveries we can only assume the following:

Vegetation – trees, low-level brush or plants, perhaps some kind of ancient grass – was indeed present in the interior of the Navajo sand-sea, particularly in its upper part, not only in-around playas, springs, and wet horizons, but also in the dune-sand, albeit limited to moist locations of still unidentified origin.

A Lifeless Desert?

Tamsin McCormick Phd.

The Navajo Sandstone is a classic aeolian deposit, with massive cross-bedding that might suggest a Sahara-like desert with towering, migrating dry sand dunes dominating the Jurassic landscape. On closer inspection, the unit reveals features that point to a variety of environments, including finer-bedded fluvial deposits, limestone lenses that are the mineral remnants of playas or saline swamps and large-scale, swirling lines suggesting soft-sediment slumping in water-saturated sands. Whereas I had previously recognized such diversity of textures within the Navajo, it was during my field trips to examine these strange tube-like features ("burrow casts") with Georges Odier that my interest was truly piqued.

Trace fossils of burrowing invertebrates (?) are not uncommon on bedding surfaces in many of the Paleozoic layers, but 'tunnels' usually have diameters of 5mm or less. However, nowhere else can I recall having seen such abundant and large tube-like structures as those that are eroding out of these bioturbated zones in the Navajo. One has to ask if they may be related to root material.

Evidence of vegetative matter is present in many layers exposed on the Colorado Plateau. Spectacular fossilized wood remnants of large trees are found in Permian and Triassic units (e.g. at Petrified Forest National Park). It is not uncommon to find bleached branching stringers in dark red siltstones of the Jurassic Kayenta, Triassic Chinle, and Permian Cutler layers, where the presence of organic matter has likely caused the red iron oxide to revert to the more water-soluble "reduced" iron and be carried away, dissolved in ground water. Elsewhere, in near-shore Pennsylvanian limestone beds or even in the core of these bleached ribbons, fossil roots may be preserved as solid cylinders of dark gray chert. In coarser layers of the Permian Organ Rock Formation, dark, irregularly-shaped calcareous stringers are seen in the host siltstone and have been interpreted to represent root fossils (rhizoliths) occurring in paleosols (Stanesco et al., 2000). Analogous features in the Navajo Sandstone may be similarly interpreted as rhizoliths.

The fact that Georges Odier really impressed upon me was the lateral extent of these irregular cylinders ("burrow casts") that were eroding out of the Navajo Sandstone. The mere abundance of these features demands further examination and contemplation. Backcountry campers are even using these erosional remnants to build fire-rings!

Having seen evidence of vegetative matter throughout the stratigraphic column, I immediately had to consider that these were in fact somehow shadows of roots or tree trunks, rather than burrow casts as Georges was proposing. The diameter alone made them atypical of any other trace fossil burrows or tracks I had seen.

On my first trip to the field with Georges, I observed numerous irregularly-shaped, dark-colored, calcareous bodies amongst the "burrow casts". It was immediately clear that these were geometrically, texturally and compositionally distinct from the larger, more abundant sandstone bodies that littered the flat Navajo surfaces nearby. At least two distinct features are present – i) a set of irregular, dark, carbonate-rich, resistant stringers, and ii) far more common, the rounded, branching cyclinders of resistant sandstone, showing a wide range of lengths and diameters, otherwise texturally and color-wise indistinct from the host sandstone. What would make these cylinders erode out of the matrix sandstone? It is likely that a more porous sand, such as might occur if a burrow-tube is filled with less compacted sand than the surroundings, will have a higher concentration of cementing material (in this case probably calcium carbonate). If these were in fact tubes, what could have made them, if it were not colonies of fairly large burrowing organisms (e.g. mammals)?

Stanesco, J. D., R. F. Dubiel and J. E. Huntoon (2000) Depositional Environments and Paleotectonics of the Organ Rock Formation. In "Geology of Utah's Parks and Monuments", D.A.Sprinkel, T. C. Chidsey, Jr., and P. B. Anderson, (eds); Utah Geol. Assoc. Publ. 28; p. 591-605.

Tamsin McCormick Phd, is the geologist for Plateau Restoration, Moab, Utah, and a member of the original investigative team.

Trunk of a very large tree that broke through the limestone bottom of a playa. The first indication that trees were growing in dune-sand prior to the playa's formation.

Piece of a large tree lying around the edge of an eroding playa, Upper Navajo Sandstone. These types of tree are currently thought to be ancient conifers. Notice similarity with contemporary dead desert trees such as junipers and pinons.

A fairly large tree still embedded in the limestone of a playa, Upper Navajo Sandstone. Its location however is not on the periphery of the playa but somewhere around 100 yards inside of it.

Typical chunks of trees lying around or inside playas. Notice dark color due to the absorption by the roots of silicate located in dune-sand, not in the playa's limestone.

Colin Egan and Tamsin McCormick examining a tree still embedded in dune-sand. These trees are the first ones uncovered outside playas and springs.

De-watering pipes in the Early-Middle Jurassic Sand-Sea

These pipes are unknown to most geologists, including the ones familiar with this region. Most are damaged by ancient and recent erosion. When undamaged, they offer a puzzling appearance: cylindrical towers, some of fairly large size, standing upright on top or sides of mesas, or 'sticking out' of slickrock. So puzzling in fact, that until fairly recently nobody knew what they really were. In the Moab region, only one (in the Dewey Bridge deposit - early Middle Jurassic) must have been seen by some geologists, as it stands fairly close to a well-traveled back-country road. In recent years another one was uncovered, but due to its location and difficulty separating it from the face of a small cliff, it is practically unknown to anyone. This particular pipe is the tallest uncovered to date in the Moab region. Still embedded into the face of the cliff it appears to begin at the end of the Navajo Sandstone deposition (?), then goes through the Dewey Bridge deposit and partially into the Entrada Sandstone. It was shown to me in May 2005 by Don Rasmussen, a senior geologist from Longmont, Colorado. Don was the first person to explain to me their origin. Other geologists I had contacted earlier didn't know what they were, nor seemed to be interested to find out.

Don Rasmussen's explanations were to open an entirely new window upon the paleo-climate of the Navajo Sandstone, especially its upper part, then later, upon the poorly known climate of the Moab-Entrada Tongue of the Entrada Sandstone – the end of the Middle Jurassic in southeastern Utah. Now informed, I then realized that I had seen several of these pipes, but puzzled by their appearance, had dismissed them as strange 'erosional features', more so due to the general lack of interest for such features.

The first one I had noticed was in early 2003 during the investigation of the 'Raccoon' track site in the Upper Navajo, 50 miles east of Moab. In-around that site, clearly located in a wet horizon, were two circles, slightly less than 3 foot in diameter, 'impregnated' in the light-colored slickrock. These two circles were of reddish-orange color with gradually smaller circles somewhat similar to paper targets used in pistol or rifle competition. Strange features indeed, so strange in fact that I took pictures of them to show to fellow researchers (without raising any interest as to their origin). Unknown to me then these 'circles' were the only visible remains of de-watering pipes eradicated by erosion. The pipes themselves had disappeared leaving only their base (or part of their base) impregnated in the slickrock. The reddish-orange color being the color of the original infilling – the color of the sand or material from an upper horizon where the pipes had surfaced and 'de-watered'. An explanation still under study but that appears to be 'in the ball park'.

In April 2005 and in the Upper Navajo Sandstone, I uncovered several of these large pipes, but still standing or broken down, on a small mesa inside Canyonlands National Park, Island in the Sky District. A puzzling sight to say the least, and one of the two reasons behind Don Rasmussen's visit in May 2005 (the other being a first-hand look at the Navajo mammal burrows). Don wasn't able to see these truly amazing pipes due to limited time. Instead he showed the tallest one, mentioned earlier, that was on a route to a mammal burrow. His explanations and deep interest in the matter triggered my curiosity and in October 2005 I uncovered another set, this time very large ones, in the upper horizons of the Moab-Entrada Tongue of the Entrada Sandstone – a major discovery up-setting the conventional view of that deposit. Don's further interest in the matter was instrumental in bringing Steve Hasiotis onto the scene.

Like most geologists and related scientists Steve Hasiotis had never seen such pipes nor for that matter was even aware of them. Upon looking at them during his late November 4 days visit he immediately recognized them as such but went a step further: He uncovered many more still embedded in the face of the small mesa, or broken, lying around at its base. These pipes however were not the easily seen large ones but a vast array of smaller ones ranging in size from 3 to 12 inches in diameter, some slightly larger. Due to my lack of knowledge at that time these smaller but numerous pipes had escaped my attention. This extended discovery is a tribute to Steve Hasiotis' observation talents and professionalism.

Following that discovery I took Steve Hasiotis to three other sites for an initial interpretation of smaller but still standing structures that I had earlier interpreted as vertical burrow entrances. All of them turned out to be de-watering pipes. This in effect compounded the impact of both the larger and smaller pipes: Instead of being rare as previously thought these pipes are present in surprising numbers, at least in the upper horizons of the Navajo Sandstone – the ones most exposed in this region.

Between Steve Hasiotis' visit and publication time I managed to uncover four more sites in the Navajo containing such pipes, most of smaller diameter, with a scattering of large ones. My investigations were then stopped by bad weather. However, taken as a whole, I was able to substantiate Steve Hasiotis' speculation that these de-watering pipes were not localized and restricted to very small areas but were part of larger de-watering fields.

As examples, I extended the size of the original de-watering field first uncovered in the Canyonland National Park to a fairly large mesa located ½ mile away to the southeast. The upper part of that large mesa – at the same geologic level as the smaller one – contains numbers of pipes, small and large, most however in poor state of preservation. Following that discovery, I then speculated that the strange mounds I had earlier uncovered during the documentation of the Texaco Burrow (a Class III type), could also be de-watering pipes. Not only are they de-watering pipes but they extend over a fairly large area, that was once an extremely wet horizon. This re-assessment is of importance. It demonstrates that 'de-watering' took place at various horizons of the Navajo (at least in its upper part) and wasn't restricted to small and specific areas. Further, and of importance to field research, the general configuration of that 'Texaco De-Watering Field' now points to similar discoveries elsewhere. Until the arrival of Don Rasmussen and Steve Hasiotis on the scene, my explorations for potential pipes were haphazard and devoid of scientific foundations. Today they are specific, albeit still in their infancy. I predict that many more of these pipes and fields are going to be discovered in the immediate future, at least in the Moab region, where most of my investigations are centered.

Note: On January 4, 2006, I uncovered a very large 'de-watering field' at the same geologic level and connected to the Freckle Flat Burrow, Upper Navajo. This is the first evidence connecting 'de-watering fields' to early mammals' presence, and most likely, vegetation and various forms of invertebrates. This discovery is too limited to imply that these 'fields' and mammals were inter-connected. It does suggest however that early mammals may have been attracted by these surges of water and the vegetation that must have sprouted in-around the surface, especially in the larger 'fields'.

These 'de-watering pipes' are just that. When large water-tables below the surface became compressed, by additional horizons, water pierced through them to relieve the pressure, somewhat similar to springs. The water surged upward under tremendous pressure creating cylindrical 'wells', that later were filled with material from the surface horizon. That material then hardened to a density higher than the surrounding matrix. Later, erosion washed away the softer matrix, leaving behind these in-filled wells as standing casts, or 'towers'. The color and material of the in-filling a reflection of the surface material. The ones in the Upper Navajo are sand-filled, thus of approximately the same color as the matrix. In the ones uncovered in the Moab-Entrada, the in-filling is made of the next deposit, the Summerville Formation, a reddish-brown sedimentary material. Thus, the color and contents of the in-filling indicate at what geologic level the compressed water exited, an indication of the length of the original pipe.

This above explanation is of course preliminary and incomplete. Formal or better studies of these pipes apparently exist but at time of publication the only one I know, was made (partially or fully?) by Marjorie Chan, a specialist of the Navajo and a geologist at the University of Utah. This information came from Don Rasmussen, with a manuscript or paper by Dr. Chan to follow. Meanwhile, Steve Hasiotis has an explanation somewhere in between: He believes that the infilling, partially at least, came from the lower horizons at or somewhere near the original water table. His speculation is based on the discovery of small to very small pipes (or conduits) connecting the large pipes to the water table. In other words, the 'roots' of the large and final de-watering pipes. His preliminary investigation seems to indicate that the infilling of these small 'roots', along with some parts of the de-watering pipe itself, is made of similar material (sand) present at the lower horizons (water table). This upward process is known and appears to be scientifically solid. On the opposite however, the large pipes uncovered in the Moab-Entrada contradict that theory. Their infillings is clearly from the next upper Summerville Formation. At printing time only Steve Hasiotis, Tamsin McCormick, and Colin Egan have seen these recently discovered pipes. Due to the arrival of inclement field conditions no one however is familiar with their various versions.

These pipes indicate fairly extensive fresh water tables, lying at whatever depth under the surface. Due to the pipes location, 200 kilometers or 130 miles west of the eastern boundary of the Navajo Sandstone, most if not all the water contained in these extensive water tables must have been of pluvial and regional origin – even if some of it originated from higher ground outside the Sand-Sea. Rain water filtering through the aeolian sand until it reached a water-tight horizon, a common occurrence throughout the entire Early-Middle Jurassic Period. The base of the pipes indicates where the water had accumulated into an extensive water-table, able to pierce through several upper horizons. This in turn indicates high pluvial episodes beginning above the base of the pipes. Thus it seems, that at least the upper part of the Navajo deposit, and the upper part of the Moab-Entrada (end of the Middle Jurassic in this region) were wet to extremely wet periods, in contradiction with current and up-to-date studies of the Navajo Sandstone asserting the exact opposite.

Note: Latest investigations seem to indicate that shallow water tables must also have been present during that time. At the 'Texaco De-Watering Field' for instance, the water table that produced these numerous pipes must have been a very shallow one – no more than 2 meters or so deep – as the Texaco Burrow lies at that lower level and immediately underneath that Field. Failing other explanations, it would indicate that some kind of heavy or torrential rain took place following the burial of that burrow. Another aspect of these de-watering pipes may have been a combination of water-tables and saturation of upper horizons by extensive rainfall. This phenomenon is known in modern times. In Nordic latitudes for instance, extensive or steady rainfall can produce muskeg-like conditions on flat surfaces. Failure for the ground to absorb unusual amount of rain produces a 'regurgitating' of the water via small and roundish 'pipes'.

These de-watering pipes seem to offer an explanation for several related factors, recently uncovered in this Jurassic sand-sea. The most important is the discovery of gigantic mammal burrows in the Navajo Sandstone. This multitude of early mammals could not have survived without water or vegetation, sufficient for their number, and this, year-around. This vegetation in turn could not have grown nor flourished year-around without fairly steady pluvial episodes. This extensive low-level vegetation is represented by a number of roots (rhizoliths), today documented or partially documented in-around these mammal burrows. In the Moab-Entrada, swampy areas, wet horizons, and fairly large numbers of petrified vegetation (bushes of some kind) also indicate steady pluvial episodes, today confirmed by the discovery of large de-watering pipes in the upper horizons of that deposit.

The pluvial episodes represented by these pipes also offer an explanation for the numbers of cross-bedded dunes commonly found in the upper part of the Navajo. Each bed (lens) representing a hardening of the surface that could only have been made by either rain or high humidity. In turn, these multiple rain-hardened beds anchored these dunes in their current locations. The inter-dunes areas then filled with wind-carried dry sand until it became level with the stationary cross-bedded dunes. When this leveling occurred, the dune-making process started all over again. Dry sand combined with wind and steady dry weather do not make cross-bedded dunes, but only move them steadily forward, leaving behind an horizontal deposit known as an 'accumulation'.

They also offer an explanation for the large number of dunes 're-worked' by torrential rains and flowing water, a common sight in the Upper Navajo Sandstone. And with this amount of rain water, especially in high tropical temperatures, surface vegetation must have taken hold over fairly large areas of this sand-sea, and flourished, during that time.

These pipes are relatively new, their numbers barely known. Comparison between the Navajo and Moab-Entrada pipes shows a marked difference in preservation: The sand-filled Navajo ones are clearly better preserved than those sedimentary-filled, in the Moab-Entrada . Sedimentary material can quickly deteriorate and become crumbly when exposed to elements. Another factor is the lack of knowledge of their existence among field researchers and geologists alike. For anyone unfamiliar with their structure they can easily pass for erosional features abounding in Canyon Country.

These pipes are additional 'proxies' for paucity of hard evidence. Similar to petrified burrows, they offer an indirect view of animals, vegetation, and the climate that brought these various elements together. Today, what we know about them, is mostly restricted to visual observations and tentative projections as expressed in this sub-chapter. While these projections are within our current knowledge of these various deposits, past and future studies could further enhance or alter our perception of ancient climates within these deposits, particularly the Navajo Sandstone.

At publication time, with a limited knowledge of the issue, here is a speculative summary:

1. De-watering pipes do come in various sizes inferring that de-watering was more common than previously thought. While the large pipes were clearly made by pressure from deep underlying water tables, the smaller ones appear to be the product of pluvial water that had accumulated close to the surface. Lesser pressure in an already damp or saturated horizon equals smaller but more numerous pipes scattered over a larger surface. In turn it would indicate that very high or torrential rains took place during that given geologic period – a speculation sustained by the numbers of wet horizons and springs recently uncovered in the Navajo Sandstone.

2. Current investigations support the presence of early mammals among some of these de-watering pipes. Thus it is plausible that these damp-wet areas may have attracted these animals due to more prolific vegetation and easier burrowing into the ground.

3. Some smaller de-watering pipes are difficult to differentiate from vertical burrow entrances. It seems that the main difference between them is in the height of the structure. Early mammals, especially the smaller ones, could not have burrowed vertical burrow entrances beyond reasonable depths, a fact supported by burrow entrances connected to horizontal tunnels found in Class II & III burrows. So, it appears that fairly tall but smaller in diameter structures found among burrows are most likely de-watering pipes. This would confirm the earlier speculation that early mammals sought and lived in dampish-humid areas, where year-around vegetation had the best chance to survive and proliferate.

4. Lack of carbonate in-around these pipes (associated with limestone deposits such as playas) seems to indicate that the de-watering process was not connected to the much slower deposition of playas and carbonate mounds, both fairly common in the Navajo.

While all these pipes, playas, carbonated mounds, springs, are of water origin, only the de-watering pipes (or the areas they are located in) seem to have attracted these early mammals. No burrow, nor tracks of these animals have been uncovered to date in-around playas, carbonate mounds and springs. A mystery that needs to be addressed to understand the ecosystem of the Navajo sand-sea. Was stagnant water represented by these playas and carbonated mounds toxic to early mammals? Or perhaps the vegetation that grew around them? Today these seem to be the only plausible answers. If correct, 'oases' as the conventional center-point of survival for varieties of animals, needs to be re-assessed.

Some of the many de-watering pipes uncovered in the Moab-Entrada deposit. The color and material of the in-fillings indicate that they had de-watered in the Summerville Formation, the next and immediate upper deposit

A large and well-preserved de-watering pipe still embedded on the edge of a small mesa, Upper Navajo Sandstone.

Steve Hasiotis examining a well-preserved de-watering pipe that has recently fallen out of a small mesa, Upper Navajo Sandstone.

A large and well-preserved de-watering pipe in the Dewey Bridge Member of the Entrada Sandstone. Notice the light-colored in-filling vs. the dark-red color of the Dewey Bridge deposit.

Don Rasmussen standing next to the tallest de-watering pipe uncovered so far in the Moab region. The water-table that made that pipe was located somewhere in the lower Dewey Bridge deposit. As indicated by this picture, that pipe de-watered somewhere in the next upper Entrada Sandstone

A still standing but eroded version of de-watering pipes, Upper Navajo Sandstone. This pipe along with others around it, is located on the flattish surface of a small mesa.

A de-watering field located on top of a small mesa, with remnants of de-watering pipes still embedded in the surface, Upper Navajo Sandstone.

Another version of de-watering pipes, Upper Navajo Sandstone. This smaller pipe, with its top lying next to it, is located in a small de-watering field among several others, some smaller, some larger. These types of pipe were originally thought to be burrow entrances.

Another type of de-watering field, Upper Navajo Sandstone. This fairly large field contains dozens of smaller pipes earlier thought to be burrow entrances.

A well-preserved and very large de-watering pipe recently uncovered in the last horizons of the Moab-Entrada Tongue of the Entrada Sandstone, Upper Middle Jurassic.

Did lack of oxygen stunt the growth of early mammals?

The fossil and track record indicates that all early mammals from the Late Triassic onward were small to very small, the largest, the size of a small dog, the smallest the size of a tiny mouse, with most about medium-sized prairie dogs. This small mammal menagerie is now backed by the diameters of their burrows in both the Early and Middle Jurassic. These burrows, however, are currently only known in the Jurassic sand-sea. Its severe climate and limited food supply could also have limited the size of these creatures until the arrival of more temperate conditions. To this we add the conventional theory that carnivorous dinosaurs kept down their size down to small prey species. Whatever the case may be in this Jurassic sand-sea – and the surrounding temperate regions of the Early-Middle Jurassic - it is now almost certain that all primitive mammals were small animals by contemporary standards.

Due to the paucity of interest in early mammals their small sizes are generally attributed to the rise of the dinosaurs at the onset of the Jurassic. A convenient theory supported by the arrival of numbers of predatory Theropods on the scene. A theory largely based on the then 'rare and small mammals' that inhabited this Sand-Sea. What happened to the other mammals that inhabited the huge temperate regions of the Pangaea Continent is seldom if ever discussed in paleontological literature. In these regions, ample vegetation and hospitable temperatures, not to mention forests and covers, offering protection against most predatory Theropods, body growth should not have been hampered, at least for some of these species. As the conventional theory goes, early mammals only began to increase in size after the demise of the dinosaurs, 65 million years ago.

On September 30, 2005, a new theory came to the surface, one vastly more believable than the 'dinosaurs factor'. This new theory was first published in the Science Journal, then again on National Geographic News, followed on the same date by the Associated Press. Written by Paul Falkowski of Rutgers University and colleagues, this new theory strongly suggests that oxygen must have been the 'culprit' for the early mammals' lack of growth during the Jurassic-Cretaceous period.

According to the Associated Press release, Falkowski and his colleagues measured samples of material deposited on the sea floor going back millions of years. By measuring the amount of carbon-13 in the samples they were able to estimate the amount of oxygen in the atmosphere at a particular time. They found that the air contained only about 10 percent oxygen at the time of the dinosaurs, as opposed to 21 percent today. Since the oxygen needs for mammals are six times as high as reptiles, dinosaurs could grow in size unimpeded by this lower level of oxygen, while the early mammals could not. The researchers concluded that the rise of oxygen "almost certainly contributed to evolution of large animals". Mammals, that is, and this after the demise of the dinosaurs, 65 mya, where the oxygen levels had gradually increased from 10 percent to 18 percent.

Following the publication of this theory in the Science Journal, etc., I contacted Paul Falkowski asking for the Abstract or Manuscript normally attached to such discovery. They were none (to date), and later asked Falkowski, or one of his colleagues, to write a side-bar since their discovery supports ours (and vice-versa). At publication time I had not received an answer, thus had to write this brief and somewhat superficial summary of their very interesting discovery.

In essence, it now appears that lower levels of oxygen were at least a contributing factor in the lack of body growth, or very slow body growth, among early mammals. However, should this theory be applied equally from the onset of the Jurassic to the beginning of the Cenozoic, body growth should be equivalent to the increase in oxygen levels. This is not the case according to the current fossil and track record. Increase in body sizes, along with the beginning of the diversification of the Mammal Class, only took place somewhere between 50 to 40 million years ago. This is not by any means a rejection of the oxygen theory. On the contrary. It is a first class explanation for a phenomenon that until today had plagued many researchers, resulting in convenient answers centered over the then dominant dinosaurs, particularly the fascination with Theropods. Robert Asher, curator of mammals at the Berlin Museum of Natural History, perhaps best summarizes this discovery: "No magic bullet, like most other issues, there are a number of causative factors involved, including chance".

At publication time Robert Asher did not know about the discovery of these multitudinous mammal burrows in the Early Jurassic of Utah, nor for that matter did Paul Falkowski and his colleagues, who only learned about their existence from my recent communications. These multitudinous burrows, their size, diameters of tunnels, colonial lifestyles, and the multitude of still unknown early mammals that inhabited them, must also have been a 'causative factor' in the lack of body growth among our Mesozoic ancestors. We applaud Paul Falkowski and his team for their contribution. More so since their discovery up-grades the determination of complex issues, such as this one, from 'one-dimensional' to an inter-disciplinary level.

We are still a long way off from a multi-pronged explanation of our ancestors' lack of body growth, but this new 'oxygen factor' must now be 'factored' in any explanations. Not a vague issue but a requirement: Because they need significant concentrations of oxygen, few living mammals today can live and reproduce at elevations greater than about 14,800 feet (4,500 meters). In the Early Jurassic, lesser concentrations of oxygen apparently did not hamper early mammals reproducing at such a high rate. But did these oxygen levels hamper their body growth?

Chapter V

The 'one-dimensional' issue

On November 7, 2004, at the Geological Society of America (GSA) annual meeting in Denver, Colorado, another page in the history of paleontology was turned, albeit unknown to most attendees. On a Poster open to both scientists and the invited public was the entire text of Abstract # 77249, the first scientific evidence that the methodology used in most paleontological determinations was in great need of review. This Poster, nor the Abstract, attracted much attention outside of a handful of specialists who were at least aware of its possible implications. What this brief introductory Abstract describes are not the skeletal remains of our Jurassic ancestors, nor their fossil tracks, but the multitude of burrows in which they lived during that Period. Burrows totally unknown to the scientific community until then.

Instead of being rare, nocturnal, and reptilian, our ancestors were in fact inhabiting the Jurassic by the millions, and right at its very onset. A failure on the part of conventional paleontology, similar if not identical to the asteroid impact affair of the 1980's. Conventional paleontology had asserted if not imposed upon other sciences a scarcity of early mammals solely based on the paucity of fossils and tracks. Professionally correct, but scientifically wanting. In other words, these scarce early mammals were 'one-dimensional', thus the failure, and the issue at hand.

Had conventional paleontology, particularly its higher ranks, progressed beyond the limitations of its own methodology, as have other sciences like the very similar paleoanthropology, this failure would have been avoided to the benefit of all. In this world of arcane studies it's difficult for an outsider to pin down the under-currents that mold evidence into solid assumptions, or the opposite, into pre-conceived ideas. Fortunately, this discovery is opening cracks in this otherwise monolithic field permitting to take a look at the origin of this failure. Beyond intercine jealousy between disciplines, lack of curiosity and imagination commonly found in close-knit circles, one can point to the narrowness of one discipline as the single factor behind this failure. Projections and assertions solely based on the paucity of fossils and tracks could only have produced rare and 'one-dimensional' mammals. And did.

This 'one-dimensional' approach is substantiated not just via field reviews of past interpretations, but queries into paleontological circles. When informed about these gigantic burrows replies were invariably the same: Skepticism, impossible, or undocumented allegations. Yet, when asked, none of these conventional paleontologists were able to describe them. When pressed to explain this discrepancy we were either informed that such burrows do not exist in the paleontological-geological record, or must have been made by termites, toads, crayfishes, lungfishes, wasps, etc., anything but mammals since they were rare to non-existent in the Jurassic. Dogmatism or incompetence?

It would be unfair however to lay the blame for this failure solely at the foot of the paleontological community. We the public, along with commercial-tourist entities and educational institutions, should share part of the burden. During the 5 years that led to this discovery, we came across an undeniable bias in favor of anything dinosaurian along with a clear lack of interest, if not rejection, of our 'rat-like' ancestors. The commercialization of the dinosaurs today has reached almost every level of our society and has permeated paleontology and other related scientific endeavors that largely depend on private or semi-private funding for their studies. In other words, Dinosaurs 'sell', not 'Rats'. To this we add a certain malaise among many when these somewhat repugnant rat-like creatures are mentioned as our ancestors. While Dinosaur Museums per square-mile continue their vigorous growth, a lonely 'Mammal Museum' is currently inconceivable, even at the national level. Especially a 'rat' one.

As one among others who had to endure this scientifically primitive methodology here is a personal experience demonstrating its depth even among highly respected individuals. I do not desire to contest anyone's reputation only to point to the type of determinations that need to be rectified since other researchers rely on them for their own studies. I chose this experience among many others because it is easy to understand, and since it took place a few years ago it shouldn't ruffle too many feathers.

I am speaking here about the somewhat famous debate over pterosaur vs. crocodilian tracks that took place during and after the investigation of the old 'Two Slabs site', earlier described in this book. A debate among 'giants' of paleontology and ichnology described in Lockley & Hunt 'Dinosaurs Tracks', published in 1995, and in F. Barnes, 'Dinosaur Tracks and Trackers' published in 1997. A debate that lasted over 10 years, thus a worthy example. The crocodilian determination was supported by no less than Guiseppe Leonardi (the famed ichnologist who, in 1980-81, coined the name Brasilichnium for Early Jurassic mammaloid tracks uncovered in Brazil) then by the no less famous Paul Olsen and Kevin Padian, a combination of specialists almost impossible to challenge. Among these giants, Martin Lockley was the only one who rejected these two interpretations for a mammalian one.

In late 2002, due to my interest in early mammals, I endeavored to check where these two slabs came from (one slab, originally). There, it only took me a short time to realize that the crocodilian determination was amiss. Not the track, the geology. At the site, and the area around it, there are no traces whatsoever of a lake, river, swamp, or water horizon that could have supported this crocodilian - known as the fossil Protosuchus, a small crocodile almost identical to the garden variety one's find in his or hers Florida backyard. The only evidence supporting this crocodile being the thin and wet horizon in which these tracks, including the mammal ones, were imprinted. A thin pluvial horizon common in the Navajo Sandstone.

During the 'debate' an explanation was inserted to support the presence of this 'crocodylomorph': "Various crocodiles are known to have existed at this time, and varieties with terrestrial adaptations even frequented desert environments". An euphemism solely based on crocodile fossils recently uncovered at the western boundary of this gigantic sand-sea. At or near the shores of the Pacific Ocean. Either sea water, or fresh water connected to the sea. Flowing or not fresh water known to have existed during the deposition of the Wingate Sandstone and the early part of the Navajo Sandstone. However, no crocodile fossils have ever been found at any distances inland from the sea. Crocodiles and turtles are living fossils dating back to the early Triassic, and all crocodile remains from that period onward are connected to aquatic or semi-aquatic lifestyles. Semi-aquatic because all crocodiles are and were semi-terrestrial – but only a very short distance from water without which they cannot survive. Therefore, 'terrestrial adaptations' is nothing more than a sleigh of hand, and 'some of these terrestrial varieties even frequented desert environments' nothing more

136

than a convenient proposition to support an implausible crocodile in the middle of the Navajo sand-sea. Crocodiles are and were present around the boundaries of all 'desert environments', like Australia and Africa, as long as water is present to sustain their semi-aquatic lifestyles – and as long as prey species are available in-around this environment – arid or not.

Armed with this knowledge I then did what this trio of ichnologists should have done: I checked zoological and biological sources to find out how this aquatic and ectothermic animal could have possibly survived in such a sand sea 320 miles from the nearest water (never mind getting there in the first place). I even checked the very details of the fairly recent discovery of Glacial Age crocodiles still living in a large fresh water pond (actually a spring) inside a desolate canyon of the Namibian Desert of southern Africa. The sum of it all, today easily accessible via the high-speed Internet, including water velocity, temperatures, depth, death-time by desiccation when water levels are low and water flow is sluggish, nutritional-survival-behavior and mating aspects, etc., of these well-known reptiles today eliminate any possibility that the trackmaker was a crocodilian.

In 1984, even in 1990, more so prior to these dates, access to related sciences was a lengthy and many a time frustrating experience. Nowadays, interested researchers have even the most complex scientific information at the tip of their fingers. Under the crocodile label for instance, one can find over 860,000 sites on the Internet depicting all aspects of crocodilian life from the Permian to the Present. Information that can be rapidly cross-checked with related sciences. This modern form of research however does not vindicate these earlier determinations solely based on the configuration of the tracks. Instead, they demonstrate a lack of sophisticated resolution toward complex problems then posed by these tracks. No 'Darwins' here, only professionals with a limited view of related sciences, the very problem of the 'one-dimensional' methodology.

But the above affair goes deeper than tracks. Assuming that the above tracks were irrevocably crocodilian, the conventional geology of that area, perhaps the entire region, would have to be re-written to match the habitat of this semi-aquatic species. In the past, geology was supreme and any animals known from their fossils or tracks had to be 'adjusted' to match the conventional geological record - like dinosaurs, early mammals, and insects, including vegetation and water sources were, and still are to a great degree, in conventional paleontology. Today, it's the other way around, geology and its related disciplines have to match the presence and numbers of these vertebrate and invertebrate species, many of them only known via their tracks and burrows. Until recently, only fossil tracks were deemed to be of scientific value among conventional paleontologists. Even then, many of them rejected the validity of these tracks as a denominator, or complimentary denominator, for projections of vertebrate species within a given region, or for that matter, the entire Pangaea Continent. For them, only body fossils were secure enough to make complex and far-reaching projections. Or, to be exact, keep these projections within the narrow boundaries of paleontology, a jealously-guarded supremacy over related sciences. Or, to push the point even further, either a disregard for other sciences or an unwillingness to extend studies beyond their own. A professional attitude but not a scientific one. Why such a lack of ranging curiosity and readiness to learn?

The answer lies in the internal structuring of the paleontological-ichnological community. Tradition dies hard, and some conventional ideas unsupported by fact can have the longest and most resolute spans. At the top are the 'conventionals', generally older, most of them in high educational positions, thus able to impose upon the lesser ones, and the public, a rigorous and conventional view of paleontology. Among them are the so-called 'progressives', but they are few in numbers, their voice seldom heard. This somewhat fossilized hierarchical structure permeates the ranks of this discipline. In essence, career advancement and access to more important or lucrative positions depend largely on adherence to conventional formulas. And when one gets 'there', the individual close-rank with the 'conventionals', a kind of closely-knit society whose survival depends not on new ideas and personal achievements but compliance with the safe, especially the uncontroversial. Although this attitude has been around for years, it must have been re-enforced by the asteroid fiasco that hit both geology and paleontology hard in the face in the early 1980s. Until that date the 'conventionals' had asserted, again based on their 'one-dimensional' methodology, that the dinosaurs extinction was due to volcanic activity, an activity they could not prove, when in fact it was due, in large part at least, to asteroid impacts that were predicted and proven by someone outside their discipline (1).

Tracks, burrows, vegetation, de-watering pipes, wet horizons, etc., are physical proxies for lack or paucity of body fossil evidence. To these we add the abstract ones, such as biology, genetics, statistics, etc. While the abstract proxies are the domain of related sciences, the physical ones are not. They are directly connected to paleontology, more so Ichnology, its sub-discipline. Ichnology, as the name implies, is the study of fossil tracks and nothing else. But tracks today are only one proxy among others. So who is going to take care of the others? Paleontology? Geology? Sedimentology? Botany? Field researchers? etc., Who? The place were they really belong is under Ichnology, the only sub-discipline dealing with proxies, in that case, tracks. This function would not necessarily entail in-depth studies of various proxies, a matter reserved to specific sciences, but would provide an initial umbrella for field researches. Something that is lacking today, resulting in scattered and ill-connected approaches exemplified by the past and current studies of the Navajo Sandstone.

Ichnology is a relatively new science still unrecognized by too many. By widening its wings to include other forms of 'proxies', like burrows, vegetation, etc., Ichnology stands to increase its academic acceptance beyond the narrow field of track morphology. One person who can lead and achieve this worthwhile objective is Martin Lockley; a man of considerable talents with an extremely rare Professorship in Ichnology. We can only hope he will give it some thought, or pass on such possibility to similarly talented individuals.

Among the few paleontologists-geologists who have recently become vocal in support of proxies as related denominators for complex projections I will note Steve Hasiotis, whose studies of proxies such as invertebrate and vertebrates burrows are today a valuable addition to our knowledge of ancient animals. I also will note Don Rasmussen who was instrumental in inserting another form of proxies – de-watering pipes – into the equation. On the abstract side of the proxies, Colin Egan and earlier, Gerard Demarcq, should be noted for both their contributions – and rejection of 'one-dimensional' propositions. There are others who under one form or another are realizing that proxies are only an extension of scientific knowledge, not unwelcome intrusions in one's specific field.

Paleontology, like any other sciences, is a continually evolving process. But in Paleontology this process is a very slow one due to the narrowness of its field. Study of ancient skeletal remains is both exhilarating and limiting. Exhilarating, because we need such lengthy and difficult studies to 'anchor' our evolving knowledge of ancient times, from fauna to geologic-climatic-botanic conditions. Limiting, because skeletal remains by themselves cannot project population density on a larger scale due to the vagary of fossil discoveries. To proceed in this avenue without supportive evidence – proxies – is fraught with dangers, as exemplified by the recent discoveries of gigantic burrows and unknown species of early mammals.

The same applies to Ichnology. Track morphologies cannot by themselves offer a fairly accurate vision of the trackmakers. An anatomical vision of the trackmaker, yes, but not an inter-connection with other animal species and their habitat. To project this most important aspect of the trackmaker one has to add immediately connected proxies and/or near-by ones. For instance, small tridactyl tracks inside or very close to mammal burrows are potentially more than tracks solely attributed to infant or not Theropods. These tracks could also have been made by early birds, 'dino-birds' or bird-like dinosaurs, either feeding on small mammals or being prey themselves – inferring some flight capability or high and agile ground-speed. Vice-versa, lack of mammal tracks and burrows in-around playas infers that the shallow and stagnant water of these ephemeral lakes was most likely toxic to mammal species.

In the Navajo Sand-Sea most if not all projections come from related proxies, since the body fossil record is next to non-existent. But, to be accurate, these proxies must 'inter-connect' with each other. A series of proxies whose 'mass' produces a fairly accurate vision of each element. Should one, or more, of these elements do not 'inter-connect' with the others, determinations are either flawed or falsified. In essence, the Navajo Sandstone is a superb laboratory where past speculations or assertions solely based on 'one-dimensional' methodologies must now be verified against ever increasing numbers of proxies, physical and abstract.

This is the tenure of this chapter, and in general, the tenure of this book. Nowadays, propositions and challenges to be taken seriously should include a lot more than 'one-dimensional' evidence. These gigantic burrows for instance are not proprietary to paleontology alone but under a fairly wide array of inter-disciplinary research. Same applies to vegetation, de-watering pipes, playas, climate, and other proxies. Taken as whole, these various proxies do stretch the intellect. But in doing so they also spearhead an up-grading of one's discipline to inter-disciplinary standards. For the benefit of the scientist, and the benefit of students and other researchers who rely on such expertise for their own studies.

Note: For anyone interested in this sad chapter of science I strongly recommend 'Night Comes to the Cretaceous' – Dinosaurs extinction and the transformation of modern geology, 1999, by James L. Powell. Publisher, W.H. Freeman & Co, New York. This highly informative scientific book written by an eminent geologist, details not only the amazing steps that led to this outstanding discovery but the professional incompetence of most conventional geologists at that time (including some of the most respected in America)

Another in the same vein but more recent is the 'Clovis Point Affair'. In spite of increasing and overwhelming evidence to the contrary, conventional archeologists had rejected any notions that American Indians had inhabited North America prior to 13,000 years BC. Thanks to the efforts of a handful of devoted and sophisticated American archeologists, the date of their arrival has now been pushed back to 20,000 years.

Side-bar

My friends the Paleontologists

I have had, and still have the pleasure to work with many paleontologists, including ichnologists and geologists. My friends, because without them this book could not have been written. My friends too, because I wish to dispel any notion of enmity or discord toward any of them, something that could be construed from my sometime sharp criticisms.

Paleontologists, Ichnologists, Geologists, and Field Researchers share one thing in common, a love of Nature and the challenges presented by the Ancient World. We are one family, with our own loves, differences, and squabbles. We also share similar lifestyles and future. To many, particularly the young, being a Paleontologist or say, an Ichnologist, has an aura of romance coupled with exotic locations and cinematic visions of Indiana Jones. Reality quickly dispels such visions. For one, there are very few professional Paleontologists in the world, even less Ichnologists, and semi-professional Field Researchers are still a rare breed. Romance and the sheer excitement of exploring the ancient world seldom insure lifestyles comparable to the level of studies. These professions are the poorest paid in the scientific world, if and when one can even locate a fairly secure position.

The motivation to enter, more so remain, in these arcane fields is 'the thrill to discover the unknown' as Frank DeCourten succinctly wrote in his now famous book, 'Dinosaurs of Utah' (1998). Another motivation is to belong to an extremely small 'club' of dedicated individuals, more interested in the pursuit of intellectual and scientific matters than run-of-the-mill financial rewards. The reason for the 'bonding' that unites these various but interconnected fields. To be sure we argue and squabble between ourselves, but this is only a reflection of the intellectual ambiance that surrounds this small community.

This intellectual ambiance is for the most our sole reward for spending months and years breaking our back under a Sahara sun or freezing our butt in Antarctica. But this ambiance is a very appealing to one in today's materialistic society. It unites professionals and amateurs in endeavors where few wish to tread. It also unites the young and the old, the attractive and the lesser endowed, and the better educated with the wannabes. No Indiana Jones nor Neolithic Raquel Welch, only individuals sweating it out (or freezing it out) under the banner of 'compassionate science'.

The lack of interest or progress in the understanding of our Jurassic ancestors, deplored in this book, is not the result of academic inferiority among paleontologists, only a reflection of their few numbers, and a scarcity of professional opportunities and funding. In the US today Vertebrate Paleontology is largely dominated by academics who are not professional paleontologists. This domination by outsiders who claim to be paleontologists, ichnologists, etc., when in fact they are only versed in these disciplines like any other interested professionals, has given lieu to a somewhat loose structure where personal and inflated views of one's field, too often override the professional aspects of paleontology. In turn, this dominance provides a 'scientific blanket' for numbers of individuals who can then dismiss, at will, speculations, determinations, and discoveries stemming from outside disciplines. This aspect is largely responsible for the 'one-dimensional methodology' and 'conventional assertions' that in the past afflicted, and still today are afflicting Vertebrate Paleontology. Paleontology is a unique discipline requiring academic preparation in basic sciences and mathematics. As such it should lead this field, not play second-fiddle to anyone as it does today. The reason for inserting this side-bar.

Chapter VI

Deliberations

The following speculations and projections, some pertinent some abstract, are within the boundary of this book. Without these discussions the discoveries described in this document would remain 'inert issues' in many scientific circles. By raising questions that can no longer be ignored a dialog between various disciplines will hopefully develop for the benefit of everyone – a dialog that has been sadly missing in the past. One of these disciplines is the use of 'proxies' to interpret or complement difficult issues. Until recently fossil tracks were the sole 'proxies' accepted in conventional scientific circles, although barely or reluctantly in many quarters. Ichnology is the only scientific field dealing with 'proxies', fossil tracks. Under the current circumstances, it would seem appropriate that Ichnology extends its scientific field beyond fossil tracks to include other 'proxies', particularly the study of burrows and related factors. It would enhance the scientific relevance of that relatively new science and open the door to more practitioners, particularly in field research.

Are mammal burrows present elsewhere in the Pangaea Continent ?

So far we don't know, if we take exception to the tunnel-casts uncovered in the upper part of the Cretaceous of Montana by Anthony Martin in 2002. These casts appear to be of mammal origin, but still today haven't received the proper documentation to make them so. The only other exception are burrows recently documented in the Salt Wash Member of the Morrison Formation (Late Jurassic) by Steve Hasiotis, Robert Wellner, Anthony Martin and Timothy Demko (and published in Ichnos, 2004). These burrows are interpreted as burrow systems of fossorial mammals, and are located near the Henry Mountains in southern Utah, roughly 100 miles from Moab (as the crow flies). As per this date, burrows similar or identical to the ones in the Navajo Sandstone in-around the Moab region have been uncovered all the way to Page, Arizona, 200 miles away, and about 50 miles eastward from Moab, all in the Navajo deposit. To this we add the still undocumented burrows in the Moab-Entrada Tongue of the Entrada Sandstone (upper Middle Jurassic) and three sites in the lower Kayenta Formation. However, these burrows are also in the Moab region, so are irrelevant to demographic projections.

The above however only reflects my own investigations, severely limited by distances, time, and financial constraints. Since, to date, I am the only one searching for these burrows it would be arbitrary and inaccurate to pre-suppose they don't exist outside the range of my investigations. When, in late 2004, I went to the Page area looking for more burrows, I went there 'blind', not knowing what to expect. Yet, as soon as I spotted geologic or other indications of their presence I uncovered them, and this, not only in Page, Arizona, but along the way while still in Utah. From that relatively brief and superficial investigation I nevertheless concluded that these burrows can be found anywhere in the Navajo deposit. A subsequent, and also relatively brief exploration of the San Raphael Reef (100 miles west of Moab) then again, the Poison Creek Canyon close to Hanksville, Utah, resulted in similar discoveries. So, as a whole, there is little doubt that multitude of early mammals did inhabit this entire Early Jurassic sand-sea. The only question, were they present beyond it?

Until we look for and uncover the same burrows elsewhere the answer lies both in logic and cross-disciplines. Without going into the complex issues of scientific projections, logic by itself suggests that if this multitude of primitive mammals could live and thrive in such an inhospitable sand-sea they must have lived and thrived in greater numbers in the more temperate and hospitable regions of the Pangaea Continent.

Until it began to break-up at the end of the Jurassic, Pangaea was a gigantic Supercontinent made of all the continents and islands of the world. During that time only two sand-seas are known from the geological record. The larger one that straddled South America and South Africa, the smaller one located on the western part of North America (the Navajo one, etc.). Geographically, these two sand-seas were only a fraction of this Supercontinent. The rest was temperate and hospitable by comparison.

All early mammals came from the fairly lush upper part of the Triassic, thus for them to voluntarily seek harsh, arid conditions over the huge temperate regions of the Pangaea Continent does not seem plausible. What is plausible, in fact almost certain, is that some of these species simply adapted to arid conditions solely for survival reasons, adaptation or death, as some species have done in contemporary times. Whether this adaptation resulted in physical or lifestyles very different from their brethren in temperate regions is unknown until we uncover their complete remains for comparison.

However, should similar burrows be uncovered in these temperate regions, it would indicate a fossorial lifestyle for most if not all these primitive mammals, thus physical or other differentiation between them would be relatively small. Using the current fossil record, albeit scarce and scattered, this appears to be true, at least in the Jurassic.

Based upon my experience and fellow researchers' studies, I predict that mammal burrows will be found everywhere in the Jurassic and beyond. I also predict that their detection is going to be a lot harder than the easily explored expanses of the Navajo Sandstone. In temperate regions preservation of these burrows is an iffy proposition due to the variations in soils deposition, vegetation, water, sediments, fluvial erosion, etc., as already experienced in the Kayenta Formation. Lastly, I also predict that, so early in the Jurassic, the main mammal lineage must have been somewhat 'monolithic', that is much less evolved than the fairly large variety present and partially documented in the Late Jurassic. While this is already known, or at least perceived, the discovery of these burrows seems to indicate a slower but somewhat homogenous evolution dictated by social and fossorial lifestyles.

Adaptation & Migrations

All early mammals came from the Upper Triassic, including Oligokyphus a lesser advanced type. In the Jurassic their evolution from Triassic stocks is poorly known due to the paucity of the fossil record. All we really know is the contiguous presence of at least three sub-orders until the end of the Early Jurassic (the upper Navajo Sandstone in Utah). After that somewhat fuzzy boundary, only two sub-orders apparently survived until the beginning of the Cretaceous, then only one – the Morganucodonts – to the end of the Cretaceous. This, solely based on the fossil record that has proved to be unreliable, following the recent discoveries of other types of mammals and the discovery of these multitudinous mammal burrows, whose makers and inhabitants are still unknown to date. Whether or not these burrowers belong to these three sub-orders is conjectural at this date. They may well belong to another sub-order, still unknown in the fossil record. The discovery of Fruitafossor in 2005 supports that possibility. Whatever the case, when these animals crossed the Triassic-Jurassic boundary they left a hospitable period (upper Triassic) and entered a similar one, except for the ones who entered the two geographically very small Jurassic Sand-Seas.

Therefore, only a very small fraction of the early mammals' total population had entered and managed to survive in these Sand-Seas. The others, the overwhelming majority, had entered and remained in, a hospitable Jurassic somewhat similar to the one they left behind. Yet, still today, we base our view of these early mammals mainly on the various species that lived for 22 million years or more in this geographically small North American sand-sea. Reason being that most fossils and tracks of these animals have been uncovered to date in this sand-sea, including the gigantic burrows they left behind. Uncovered there simply because large portions of this ancient sand-sea lay exposed to explorations, particularly the Navajo Sandstone. By comparison, the adjacent but more temperate regions of the then Pangaea Continent are a lot harder to explore resulting in only a small fraction of the current fossil and track record. In southeastern Utah the difference in exploratory difficulty can be easily seen between the somewhat easy to access and explore Navajo deposit and the lower Kayenta Formation, a deposit somewhat resembling the regions outside that Sand-Sea.

Thus the pertinent question: What happened to the multitude of early mammals that lived and evolved in the more hospitable regions?

From the Late Jurassic record, anywhere, their fossils indicate dozens of various species, but all 'rat-like' and generally small. A range in size from badgers to moles corroborated by the diameters of these newly uncovered Early Jurassic burrows. However, we also know from recent fossil and track discoveries that at least a few of them were larger, some roughly the size of a small dog. A size corroborated by the Tritylodontid Kayentatherium (Middle Early Jurassic) and the recent discovery of Repenomamus giganticus in the Early Cretaceous. Assuming that the current Late Jurassic record is in the ball park, it infers that the early mammals that lived in both hospitable and inhospitable regions of the Pangaea Continent had evolved from the Early Jurassic onward in approximately the same manner, size, and methods of survival.

However, living for 22 million years or more, in an inhospitable Sand-Sea and living in fairly lush hospitable regions are two different things, the first one demanding a fairly severe adaptation, the other no more than common lifestyles found today in temperate regions. These newly uncovered Early Jurassic burrows inform us that some species of early mammals had indeed adapted to the severity of that Sand-Sea, and in fact had proliferated in it until at least the end of the Navajo deposit (end of the Early Jurassic in Utah). Similar burrows uncovered in the Upper Middle Jurassic (Moab-Entrada) indicate that at least some of these Early Jurassic 'desert' species had survived through the even deadlier Middle Jurassic period (Entrada Sandstone) but so far appear to be the smaller types among them.

We now know that the upper part of the Navajo deposit was a wet to perhaps extremely wet period, from water-damaged sand dunes, multitude of rain-induced cross-beddings, vegetation, to the recent discovery of numbers of de-watering pipes. This wet period, in reverse, must have produced an equivalent amount of vegetation. This somewhat lush vegetation is confirmed by the multitude of petrified roots found in these upper Navajo burrows, in turn instigating an even more prolific mammal population – a fact corroborated by the gigantic number of many of these burrows at that particular geologic level. From this current and newer picture of the Early-Middle Jurassic of Utah, it appears that either a 'mass extinction' (of sort), or a 'mass migration' (of sort), took place at the end of the Early Jurassic (Navajo Sandstone). In plain English, upon the arrival of the deadlier Middle Jurassic period (Entrada Sandstone) most or many of these 'desert' early mammals either perished or migrated to surrounding temperate regions.

Perished or migrated? Perished is easy to visualize, but migrations, especially for such small and vulnerable animals, are a lot tougher to explain. To begin, if these animals had, at the very beginning of the Jurassic, found this Sand-Sea too inhospitable to survive in, they would have either perished or migrated elsewhere. Instead, they adapted to these new conditions. After 22 million years of adaptation their physical, etc., evolution may have precluded migration to, and re-adaptation to conditions now potentially 'foreign' to them. I say may, because until we uncover fossils in these Navajo burrows we do not know their level of adaptation to desert conditions. And by inference, the evolutionary differences between them and the majority living in temperate regions.

One thing however is certain: The early mammals that lived in the temperate regions of the Pangaea Continent are, without any doubt, the main lineage of the Mammal Class. The only question remaining is: Are the mammals that lived in these Sand-Seas part of that lineage? (through migrations, or through a narrow and continent-wide evolutionary process that was not affected by living conditions).

That last question is at least partially answered. Some of these Early Jurassic 'desert' species did not die, nor migrated. They survived through the deadly Middle Jurassic (Entrada Sandstone) as their burrows confirm. These burrows, on the other hand, were made by small fossorial species that had adapted to desert conditions now for more than 40 million years. Until we uncover their skeletons we can speculate the following: After 40 million years of adaptation, it's more than likely that they remained fossorial and prey species from that later period of the Jurassic onward. Not part of the main mammal lineage – the evolutionary branch - but a 'living fossils' branch that has survived, with no major evolutionary changes, until contemporary times.

Equivalence

The first one, and by far the most influential, is a simple equation still misunderstood by many professionals.

Tracks (any tracks) = Water

Now, including its four basic components, this equation reads as follows:

Tracks = water = vegetation = invertebrates & vertebrates = predators.

This is called the terrestrial 'Cycle of Life', and it's impossible in this Food Pyramid to remove or ignore any parts of the equation without removing the whole. But in the past and still today some elements of this equation were and are still conveniently ignored in order to conform to conventional views of geologic periods. For the uninitiated here are brief explanations of each of its elements:

Tracks:
All tracks, repeat all tracks, were made and preserved in surfaces ranging from moist to extremely wet. No tracks, repeat no tracks, could have been possibly preserved in dry sand or similar dry surfaces. Their preservation is solely based on the higher lithification (density-hardness) of all and any moist-wet surfaces.

Water:
Fresh water comes in two forms: Pluvial or Fluvial, and/or a combination of both. In the gigantic Sand-Sea of the Early and Middle Jurassic all fresh water was pluvial. Except for a few and short fluvial penetrations, water that occasionally flowed from the higher but non-aeolian grounds resting at the eastern boundary of this gigantic Erg (the Uncompahgre Range). The Kayenta and Moenave Formations were fluvial or semi-fluvial 'tongues' that penetrated the aeolian Sand-Sea Springs, but only regional and intermittent deposits. Playas (ephemeral lakes), wet horizons, etc. are of pluvial origin, even the water tables that then laid at great or not depths under the surface. Thus, any tracks found in aeolian deposits indicate rain or a very humid period, or made in wet-moist surfaces of pluvial origin.

Vegetation:
Vegetation comes with water, any forms of water, even under the most periodic conditions. Type and density is directly connected (and adapted) to the amount of precipitation in a given region. As a rule, episodic rain translates into ephemeral grasses and plants,   springs and playas into trees and brush. When rain episodes are fairly constant, grasses and plants are no longer ephemeral but growing and reproducing year-around. Trees and bushes can survive long periods of drought, their deep roots reaching moist horizons continually replenished by playas, springs, and water tables. However, they cannot survive like grasses and plants when deeper and moist horizons are either depleted or not present.

Vegetation by itself is also a form of water. Most insects and some vertebrates can feed and survive on it without drinking any water year-around (although rare among vertebrates, including birds). But this aspect is somewhat irrelevant since this vegetation could not have been present without some forms of water.

The third aspect of vegetation is a lot deeper (and botanically controversial) than vague plant species conveniently supporting a vast array of insects, mammals, and herbivorous dinosaurs. In fact this third aspect is so powerful that it is challenging by itself the conventional habitats of the Early and Middle Jurassic of the Southwest:

There was no vegetation to speak-of in the conventional geologic record of the Early and Middle Jurassic of southeastern Utah. Here again we come face to face with the old method of evaluating geologic formations and their habitats: 'Show me the fossil (vegetal in this case) and I'll tell you what the habitat was'. Since there were no, or only handful of vegetal fossils in the Early and

Middle Jurassic of this region (until trees were recently uncovered in the Navajo and Kayenta Formations), the whole 40 million years period was dismissed by geologists as an 'unforgiving, water-less desolate desert'. Paleontologists then abiding by this rigid but superficial geologic record concluded that such sterile environment could only support 'Theropods eating each other'. Or perhaps Theropods getting larger and larger solely on insects, as some are still suggesting, since mammals were so 'rare and nocturnal'.

What about the tracks? tracks?, what tracks? (1) In their rush to be agreeable to each others these conventionals had 'forgotten' the tracks.

Tracks = water = vegetation, thus any professional worth his or her diploma today must assume that vegetation was present in some form or another regardless of the scarcity or lack of vegetal fossils. Now to the next step: When mammal tracks are present, as they are now throughout the entire Early-Middle Jurassic of this region, we must conform this vegetation to the foraging need and abilities of this multitude of small animals - and not the other way around as it was in conventional geology and paleontology. So what comes first, geological and habitat interpretations solely or mainly based on 'One vegetal fossil - One vote' ?, or the other way around, based on the ever increasing presence of mammals now backed by multitude of burrows?

If you believe that such matter can easily be resolved via 'negotiated compromises' between conventional scientists let's move on to two implications raised by this multitude of early mammals: Were they foraging in trees, or were they foraging on low vegetation, perhaps some kind of plant-grass unknown in the botanical record?

In the Navajo Sandstone we now know that there were two type of vegetation. Trees, most documented as conifer species (2), and large numbers of rhizoliths (roots) found in-around these burrows indicating some kind of abundant brush or low-level vegetation. Trees were first uncovered in 1997, roots in 2003 but only documented in 2004. These trees, some up to 3 ft. in diameter, have only been found in-around playas (ephemeral lakes) and springs. Roots, on the opposite, have only been found (so far) in-around burrows. These roots are not casts like burrows, but fossils. The larger ones are easily distinguished from burrows by their greenish-brown color and/or by their calcareous or siliceous infillings. In some the root itself had rotten away prior to or after burial by the next horizon and were later in-filled with fine sand or particles, but their morphology and systems are different than vertebrate or invertebrate burrows. The smaller to tiny ones are harder to locate and identify, generally the domain of specialists. Plants such as fern (tempskya), horsetail (a very common plant in the Jurassic known as equisetum), and stalked green algae (charophytes) have yet to be documented in sand-dune and in-around burrows, although they appear to be present in oasis located on the periphery of the Sand-Sea, or in oasis made by fluvial penetrations. In the upper Middle Jurassic of the Moab region (Moab Entrada), no trees but some kind of low-level bushes have recently been uncovered in fairly large numbers (3), along with roots that appear to be swamp-like vegetation.

Should we accept this above and current botanical record as representative of the Early-Middle Jurassic of this region, we have to choose between trees and low-level vegetation as the main diet, perhaps the only diet, for this multitude of mammals. I believe that some kind of plant-grass-brush is the correct answer because no trees have been uncovered so far in-around any of these burrows

- neither in the Early or Middle Jurassic. Further yet, no mammal tracks have been uncovered to date in-around these playas and springs, and this in spite of years of investigation (4).

In a kind of summary of the above paragraphs, mammals and vegetation are integral but we still don't know what type of vegetation these mammals were actually feeding on. The only thing we can surmise is that this vegetation must have been prolific and nutritious, and this, year-around. This is where we add another equation to the vegetation 'thing', not yet on the radar screen but one that will have to be addressed in the near future. It goes as follows: x-amount of vegetation foraged daily by a given mammal or herbivorous dinosaur X by x-number of mammals or herbivorous dinosaurs in a given area X by x-numbers of days foraged in that given area = x-amount of vegetation present in that given area during that time-period.

In professional circles that equation (actually several of the same type) is usually refers to as 'Range Evaluation'. Not vague equations but some of the most complex and sophisticated ones ever devised in the scientific field (5).

That equation can be 'reversed' or manipulated to produce numbers and types of animals instead of vegetation 'tonnage'. Currently, these equations only deal with contemporary animals and vegetation. In these, at least one of their components can be assessed in the field thus insuring fairly accurate projections. Can such equations be applied to prehistoric times where any of its components are still vague or conjectural?

Yes, but the resulting projections would be so statistically inferior to contemporary ones that they would only represent a 'myopic' view of the place and period. But a 'myopic view' is a lot better than no view at all - and certainly better than arbitrary ones. In defense of conventional geology-paleontology it should be noted that the use of statistical-mathematical projections - or the use of scientific components outside the boundary of these disciplines - was not even contemplated a few years ago (and still are not in many quarters).

Nowadays statistical advances backed by the almost unbelievable capabilities of computers are able to create abstract models and projections that defy the mind - and defy interpretations solely based on sparse evidence or the narrowness of a given discipline. To be sure we are not 'there' yet in paleontology-ichnology-geology but certainly on our way in that direction (6).

The number and size of these burrows now offers - for the first time - a limited but still reasonably accurate 'base' for some statistical work. However primitive this abstract work would be, and would remain for a while, it would nevertheless bring 'into focus' some aspects of these early mammals that, in the past, would have been impossible to imagine or detect via the track and fossil record. We are not talking (yet) about vegetation 'tonnage' per square kilometer, but the (statistical) presence of low-level vegetation in areas that were presumed devoid of both vegetation and animals, mammals in particular.

Invertebrates & Vertebrates:
The vegetation-insects equivalence is so routine it's not worth a discussion. On the other hand the vegetation = vertebrates is far from routine since we are dealing here with types of animals - dinosaurs or mammals, or both.

Beginning with early mammals, most paleontologists (and zoologists) still believe they were or must have been insectivores, either by design or opportunity (survival). So wide-spread an assumption, that today all early mammals are categorized as 'insect eaters'. Based on comparative anatomy - more so dentition - with somewhat similar contemporary mammals this assumption is or appears to be undeniable. Therefore without going into the complex aspects of omnivorous animals (7) - something we shall discuss in the next sub-chapter under 'Predation' - we can safely assume that vegetation = insects = presence of mammals, at least potentially.

But the presence of both carnivorous and herbivorous dinosaur species is a lot more complicated. Here we are no longer dealing with 'generic' vegetation, but specific types capable to sustain x-number of fairly large herbivorous dinosaurs, known (or not) via the fossil and track record. This translates into down-to-earth habitats where each species is equivalent to one another up and down the Food Pyramid. But in the Early-Middle Jurassic of southern Utah - unlike the mammals who thankfully left their burrows behind for us to study and drool upon - the dinosaurs only contribution to the effort is a fairly large number of Theropod tracks (predators) and a sketchy and fragmentary fossil record indicating the presence of some herbivorous dinosaurs (8). Since the track record is the only solid evidence of the dinosaurs-at-hand, it was then assumed that all these Theropods must have been 'eating each other' (not mammals because they were 'rare and nocturnal'). This cannibalistic menagerie was also convenient : It simply by-passed the vegetation question. Convenient also to by-pass herbivorous dinosaurs, the only possible prey for, at least, the medium to large Theropod species. A whimsical habitat and menagerie of course, but given the then in vogue 'one-fossil-one track-one vote' formula it would have been hard to come up with anything else. Now back to the reconstruction of realistic habitats and fauna:

Today (and again) the only real means currently at our disposal to estimate the ratio between predators and dinosaur prey species - and the equivalent vegetation to sustain such species - are statistical projections. However, unlike mammals, statistical projections are complicated by the known presence of Sauropod-like species (9). While hampered by the current scarcity of comparative studies (10), basic projections can still be achieved via computerized models and statistics. This so because the pyramidal limits and variations of the so-called Food Pyramid in contemporary terrestrial vertebrates are already established and within scientifically acceptable margins of error.

While this statistical work will take a certain time to impose itself as an additional and new mean to interpret prehistoric species and their habitats we can at least assume the following:

1. Medium to large Theropods in the Early and Middle Jurassic of Utah did not and could not have fed on small mammals as their primary diet - and therefore, to survive had to feed on herbivorous dinosaur species, whether they exist or not in the fossil and track record.

2. The ratio between predators-preys is at the bare minimum 3 to 1 in large terrestrial species. So, regardless of the prey (herbivorous dinosaurs or smaller Theropods) this extremely minimal ratio indicates that, at the very least, an equal number of large Theropods and fairly-large herbivorous dinosaur species must have lived side-by-side in that period of the Jurassic.

3. The suspicious if not incredible lack of herbivorous dinosaur tracks in the Early and Middle Jurassic

is statistically inconsistent with the current dominance of tridactyl tracks attributed to predatory Theropods. Since the ratio between predators and preys cannot be changed, nor manipulated via convenient but undocumented assertions such as extreme cannibalism, these multitudes of tridactyl tracks must be reviewed to include a similar ratio of herbivorous Ornithopods, bipedal dinosaurs that left behind similar if not identical tridactyl tracks (11).

4. Regardless of who these prey species were, vegetation had to be equivalent to their number and position within the rigid Food Pyramid. This vegetation equivalence extends to all prey species whether be herbivorous dinosaurs, mammals, or insects. Whether or not a type of animal or insect was directly involved in the survival and eventual proliferation of certain Theropods species, the size and survival of these predators were still directly connected to the size of the vegetation - the lower and fixed component of the food chain without which no terrestrial species can survive.

5. Cannibalism cannot by-pass the Food Pyramid, even under the most extreme conditions. Larger Theropods feeding on smaller ones, or each other, does not exempt vegetation. At the very bottom of this cannibalistic chain lay the terrestrial invertebrates and small vertebrates that depend on vegetation for their own survival. The smallest of the Theropod prey species (infants or bird-like dinosaur species) had to have equivalent numbers dictated by the rigid Food Pyramid, numbers that grow larger and larger from the Apex down to the preys. These large numbers of intermediate preys then again translate to even larger numbers of smaller preys, this time, invertebrates and small vertebrates that again translate into equivalent amount of vegetation, the bottom of the Food Pyramid. However, this extreme view of cannibalism is not supported by contemporary zoology. While cannibalism exists in practically all carnivorous-predacious terrestrial animals it is never a form of sustained diet, at best only a portion of it (like scavenging). Had the Theropods of the Early-Middle Jurassic only used that method of survival they could never have achieved their subsequent numbers, sizes, and varieties.

6. The presence of fairly large numbers of small dinosaur prey species needed to support a cannibalistic method of survival, particularly carnivorous ones like infant Theropods or bird-like dinosaurs, is not supported by the up-to-date track record. Tracks of infant herbivorous dinosaurs are non-existent, and tridactyl tracks that can be attributed to either infant Theropods or Ornithopods (including bird-like dinosaurs) are extremely rare in the Navajo Sandstone. So rare in fact, that tracks attributed to early mammals are numerous by comparison. Taken as a whole, the Navajo track record indicates the presence of only two dominant species: Small-medium to large tridactyl tracks attributed to Theropods or Ornithopods?, and early mammals. This in turn translates into infant Theropods (or Ornithopods) protected until adulthood in nests or similar protection, and small 'bird-like' dinosaur species whose speed and agility must have put a lot of strain on any cannibalistic notions. Whatever the case, these infants nor bird-like dinosaurs would not have been sufficient in numbers and size to sustain an adult menagerie of medium to large Theropods ranging in weight from 200 to 700 pounds. To grow fat, sassy, and larger as they did, these Theropods were feeding on something else than their young, 'jack-rabbit' dinosaurs, and intelligent mammals protected from these clumsy dinosaurs by deep and inter-connected city-like tunnels.

7. The amount of vegetation that was needed to sustain this new and more realistic menagerie of predators and preys represented by various tracks, and now burrows, must have been substantial. Not a scientific term but a lot more accurate than the presumed lifeless and unforgiving deserts of

the past.

Predators

Predatory species from higher vertebrates to the most minuscule bacteria are integral with the natural Cycle of Life on our planet. They 'Come with the Territory' and are present in even the most basic form of life, water. Every class of animals, ancient or contemporary, has its own predatory species (12). Predatory insect species are not part of the equation since dinosaur and mammal predators were the only ones that could have affected the vertebrates-vegetation equivalence during the Early-Middle Jurassic period. However in the conventional food chain of the Jurassic it is assumed that predatory dinosaurs (Theropods) were the only ones preying on other vertebrates. This assumption is largely based on physical considerations: Theropods being the only ones around with a predator's anatomy and dentition. It was also anchored in the fascination for dinosaurs and a resulting lack of interest in early mammals. As we shall see in the following sub-chapter the predator-prey relationship in the Jurassic is in need of review.

Predation

Who ate Whom?

Predation is not a pretty word, nor it is easily accepted as a method of survival. Yet it is part of the evolutionary process in practically all species, ancient or living, mammals included. So let's brush aside philosophical considerations and stick to its scientific aspects:

1. Predation can be external or internal within a class, order, family, genus, but seldom within species. External means predation outside the given animal class. Although external and internal predatory species are usually lumped together as predators the difference between them is immense: External predation (inter-class predation) is part of the rigid food chain while internal predation (class predation) is an internal mechanism to cull populations to sustainable levels within a given habitat. While these two forms of predation appear to be similar if not identical since the result is approximately the same - stable populations within the food chain - they are not.

In the conventional Jurassic we were told that Theropods were the only predators around, presumably feeding either upon themselves, rare mammals, or nearly non-existent herbivorous dinosaurs. Further, we were told that these rare mammals were either herbivorous or insect eaters - depending on whom you talk to. However no one has yet to explain how these Theropods, especially the medium to large ones, could have possibly preyed upon these burrowing mammals.

Theropods were bipedal, an evolutionary process dictated by speed, because speed was and still is a major component in the survival of most terrestrial predators. Not only bi-pedal but equipped with a unique anatomical configuration known as the metatarsal condition – three 'hinges' in the hind legs instead of two as per all mammals, a feature today only found in birds and bird-like species. But bi-pedalism is the worst possible anatomy to dig with, more so under the metatarsal condition. Some contemporary bipedal species (like penguins, swallows, petrels, etc., including kangaroos) can construct some kind of burrows ranging from shelters to fairly deep affairs. All of them use their beak to do so, except kangaroos or similar species that have small but still functional front legs. However, construction of burrows and digging for preys are two different things. Today the only predators capable to dig for burrowers are quadruped species with extremely developed shoulders, like bears or badgers. And even they have difficulties to dig for very small ground squirrels (13).

Only three years ago nobody really knew how these Jurassic mammals had managed to survive in such inhospitable habitats. Burrows were a logical guess. Today they are no longer a guess but a documented reality. Not only multitudes of them, some of gigantic sizes, but also cross-sections of them confirming the depth and horizontal spread of these tunnels - the sum of which can be summarized as follows: It would have been inconceivable for any of these medium to large Theropods to have successfully dug for these burrowers, at least as a primary sustenance.

For these Theropods the only successful way to prey upon these small mammals must have been on the surface. Here we are on sturdier ground since we can use contemporary burrowers-predators comparisons. The foremost difficulty these Theropods would have had to approach these mammals must have been striking distance. In ancient times as it does today, any approaching predators would have set off an alarm throughout the entire colony. These early mammals were vastly more intelligent than any dinosaur species, not 'easy picking' (14). This easily verifiable reality strongly suggests that only small to very small but fast and agile Theropods could have successfully hunted for them (15). The fear factor gradually diminishes as the predator size gets smaller, which in turn brings easier concealment and success. This proposition seems to be supported by the up-to-date track record of southeastern Utah - and this with far-reaching implications:

No medium to large Theropod tracks have been uncovered to date inside or around these multitudes of burrows. The only ones present so far, but in minimal numbers, are small to very small 'bird-tracks' (16).

In summary, the constant culling of prolific mammal populations necessary to insure their survival, could not have come, at least not enough, from predatory dinosaurs, large or small. It must have been internal. Thus, by inference, some Jurassic mammal species must have been predatory to their own kind either by design or opportunity. In colloquial terms, "It took an Apache to catch an Apache ".

Not a loose proposition. We had been told in no uncertain terms that our Jurassic ancestors were 'nocturnal and insectivore' when in reality, some of them were feeding on the dinosaurs themselves, albeit infants. If these newly found and carnivorous mammal species were capable to feed on, infant or not, dinosaurs their feeding upon their smaller counterparts must have been a 'piece of cake' by comparison. Rest assured that the conventionals who were caught with their pants down by such inconvenient discoveries, will soon come up with teeth configurations for Repenomamus Gigantic and Robustus solely designed for eating dinosaurs, and should they have been feeding upon their brethrens, these must have been 'mammal-like reptiles'.

Predation takes many forms. Direct (external) or indirect (internal) but hostility is also a form of predation. It deals with territorial dominance, thus survival. The territorial imperative must have been present everywhere throughout the entire Jurassic, more so in the Early-Middle of southeastern Utah, where harsh environments could only produce limited food supplies. But the final product of direct or indirect predation has, per force, remained constant throughout all prehistoric and contemporary times: the smaller the prey the more prolific it must be to survive. While smallness is by itself a form of survival high fecundity is the only insurance against constant and never-ending predation. For instance, small and defenseless species like rabbits and mice are constantly replenishing their ranks via numerous and prolific litters. In the Jurassic, decrease in sizes must also have been accompanied by an equivalent increase in fecundity. And this need for fecundity must also have been a factor in the evolutionary process to placental reproduction - a higher method of survival.

Today, early mammals are mostly known via the configuration of their teeth and jaws. Anatomical comparisons with contemporary species are slowly emerging but are still a distant second to teeth-jaws interpretations (17). Here is the latest version of the teeth-jaws interpretation published on the April 2003 issue of National Geographic ('Rise of Mammals') and written by experts, in their own words:
" The upper and lower molars of morganucodontids jawbones interlocked, letting them slice their food into pieces. That released more calories and nutrients " (R. Gore). ....   " Reptiles don't cut up their food, they grab and gulp. But these little guys were so active they had to get every calorie they could out of what they ate. The more they could process their food in their mouth, the more energy it gave them " (R. Cifelli, Sam Noble Oklahoma Museum of Natural History). "The separation of the jaw and the ear bones allowed the skulls of later mammals to expand sideways and backward - enabling mammals to develop bigger brains "(R. Gore).

This brief and technical description (out of a 37 pages article) carefully avoids any discussions or possibilities that, at least some of these Jurassic mammals, may have been omnivorous, i.e. flesh-eaters, thus potentially predatory. It did so because it was written, as most of these articles are, from the narrow point-of-view of one discipline. A bit surprising I must add, since the article deals at length with another discipline – genetics - that is currently challenging paleontology as the sole or main interpreter of pre-historic species. Surprising also, since the same paleontologists by-pass their own record.

The paleontological record speaks for itself :

In the late Triassic, the Therapsids - the direct ancestors of Jurassic mammals - were both carnivorous and herbivorous, with omnivorous species in between (like Morganucodon and Diarthrognatus). Immediately after the K-T mass extinction, the early Cenozoic, we are back with both carnivorous and herbivorous species, presumably again with omnivorous types in between. So, to dismiss carnivorous, more so omnivorous species at-of-hand in both the Jurassic and the Cretaceous - regardless of teeth and jaws configuration - is not only arbitrary but an off-the-cuff rejection of basic zoology.

The term 'carnivorous' (meaning flesh-eating) usually describes predators. But the term 'omnivorous' is an ambivalent one since plant and flesh eating can apply to both predators and prey species. For instance Man, bears, pigs, hippos (18), mice, are omnivorous - and this regardless of their dentition. Some species of rats and mice are 70 % flesh-eaters (19) with dentition similar or close enough to many Jurassic species. Not only flesh-eaters but predatory as well (capable, when the opportunity or the néed arise, to kill preys up to their own size). Rabbits, one of the most common prey animals, when under severe stress can and do eat their newly born. Thus, to use

154

dentition as the sole mean to classify ancient or modern mammals as 'vegetarians' or 'insectivores' is too one-dimensional to be an arbiter in methods of survival. The lack of preserved soft organs in skeletal remains is not an excuse for loose or arbitrary determinations: Many, if not most of these early mammals, were close to very close anatomically to similar contemporary species (20).

We can at least assume that these early mammals were 'opportunistic vegetarians' - first making use of abundant or scarce vegetation, then expanding their diet to include insects and flesh (dead or alive), when the necessity or the opportunity arose. Most likely they ate anything in their path to survive: vegetation, insects, and whatever flesh they could get their teeth in - like mice and rats do, like we do too. The Jurassic was not a romantic period for our ancestors. It was a pitiless one dictated by survival.

To survive and thrive, our ancestors must have had to cull their prolific populations themselves. This may or may not have been the case during Triassic times, but it must have been the case during the Jurassic and Cretaceous periods. The multitude of some of these burrows attests to a rapid demographic expansion that, sooner or later, in the arid Navajo deposit, would have exceeded the available food supply (vegetation).

While this proposition seems logical and well-founded into our knowledge of current mammal species, what really happened in the more temperate regions of the Pangaea Continent is conjectural. Here is a tentative speculation: Should the mammal species that lived in these temperate regions turn-out to be fossorial, then internal culling would also seem to be a fair proposition.

Living fossils?

Could it be that some small contemporary rodents are 'living fossils' with their roots in the Jurassic Period?

To jump across 165 million years of evolution and suggest a connection between vague Jurassic trackmakers-burrowers and the mouse roaming around your kitchen at night seems to stretch the imagination to the edge of science-fiction. Yet, current tracks and burrows discoveries, along with new genetic and anatomical studies, seem to indicate such possibility. M. R. Voorhies was the first scientist (to my knowledge) who, in 1975, pointed to some contemporary rodents as 'living fossils' (21). More recently, Prof. Gerard Demarcq (Paleontology Dept. University of Lyon, France) also supported such possibility based upon anatomical and genetic similarity between some ancient and modern prey species (mice-like species). However, most paleontologists are luke-warm if not antagonistic to such a proposition again based on the conventional theory that early mammals had not diversified until after the demise of the dinosaurs.

Wandering into the world of contemporary rodents is a mind-boggling proposition, even if one tries to reduce the foray to diminutive critters. For instance, there are 120 different species of North American mice alone! The objective being to isolate the smallest group possible with physical and adaptation affinities similar to very small to tiny Jurassic species we can tentatively set-aside the Deer Mouse (Cricetidae), the Desert Shrew (Soricidae), the Pocket Gopher (Geomyidae), and the Least Chipmunk (Sciuridae), a fair cross-section at least for now. But more important their tracks are well-known (in sand, mud, or snow) along with detailed studies of their feet structure. Most of these mice-like animals have front feet with only four functional digits (the fifth, actually the thumb, being vestigial) - and five functional digits in the hind feet (also larger than the front ones). Of further interest is their foot 'pads'. They vary between these four species but the following should be noted:

From the 120 different species of North American mice, about half of them fall under the general classification of 'white-footed mouse', like the common Deer Mouse. Their front feet have four toes with nails, three palm pads, and two heel pads with a vestigial thumb located near the inner heel pad. Their hind feet have five toes with nails, three palm pads, and three heel pads. This common foot structure leaves track patterns in mud and snow resembling some Jurassic tracks, like Ameghinichnus, a track type attributed to a Middle Jurassic gerbil. The Eastern Chipmunk (similar to the Least Chipmunk) tracks do resemble (as expected) the 'Squirrel' type recently uncovered in the Wingate and Navajo Sandstone. Both in size and configuration but also in the track's width. Tracks undoubtedly made by similar-sized but very ancient rodent-like mammals. While body fossils are the only way to know their actual anatomy we can at least assume a general similarity to contemporary rodents of the same size and lifestyle.

While any connections between tracks of contemporary rodents and somewhat similar prehistoric species seem to be a fanciful extrapolation at this stage, the following information is a sobering reminder that outright rejections solely based on pre-conceived ideas are not in the best interest of science. The life-like replica of Hadrocodium (the original one made by Chinese paleontologists) was not the product of the imagination. This replica was put-together using X-rays of a living field mouse similar in size to the Hadrocodium fossil, but also similar in skeletal structure. This is how the muscles, living tissues, skin, general appearance, etc., were added to the Hadrocodium skeleton, producing a fairly accurate life-like rendering of this Early Jurassic species.

In other words, as opposed to drawings, renderings, etc., of other prehistoric animals based on a great deal of imagination however educated, the Hadrocodium life-like replica is based on a contemporary living species. A species close enough in overall skeletal structure to match this fossil after 195 million years of evolution. Did this field mouse used for the replica evolve from larger and obscure ancestors in the Early Cenozoic then regressed back to its contemporary size? Or perhaps it never evolved beyond its ancestral size, and retained the same and effective methods of survival developed by its Jurassic ancestors all the way to contemporary times ?

This question needs a better answer than current and blurry assumptions of the true mammal lineage: Both today and in Jurassic times these mouse-like animals were, and still are prey species at the bottom end of the vertebrate food chain. A fact, not an assumption. Did these tiny Jurassic mouse-like species cross the K-T Extinction then, somehow, without any plausible explanations, split almost immediately into two opposite groups of preys and predators ? A bit far-fetched to say the least, yet this is the conventional explanation for the mammal lineage out of the Jurassic-Cretaceous period.

For instance here is a major discovery made by Dr. Andre Wyss (University of California at Santa Barbara) in Madagascar in 1999 (22). Diminutive teeth - no larger than the head of a pin - of a species of tiny Trebosphenidians , dated at 165 mya (Upper Middle Jurassic), put an end to the conventional theory that the true mammal lineage had solely evolved in the Northern

Hemisphere. This discovery, if nothing else, confirmed for the first time that the early mammals of the Jurassic Period were in fact a vastly more varied lot than previously thought, whether or not these trebosphenidians are part of the general Morganucodontid Group. Further yet, the Trebosphenidian Family includes both early marsupials and protoplacental mammals inferring that these tiny species were or may have been the main and true mammal lineage in the Jurassic Period.

Really? solely based on teeth configuration?

Recently discovered tracks and burrows tell a different story. The most important part is that two different types of burrowers existed side-by-side, or close to each other in the Navajo Sandstone. These two types can be identified not only by the diameters of the burrows but also by their patterns. Small-medium to large diameters, including entrances, clearly housed animals ranging in size from contemporary rats to badgers. These burrows indicate a variety of animals ranging from small to fairly large (like badgers), most being in the smaller range. However, much smaller burrows, including entrances (sometime inside larger burrows, sometime as separate colonies), indicate separate and very small species now backed by their tracks.

In the Navajo Sandstone these very small burrows were first thought to be of insect origin (23). Some may still be but most are compatible in pattern, cross-section, infilling, and entrances with the larger burrows, except smaller. In the Moab-Entrada, similar burrows were uncovered next to undeniable invertebrate ones (termites-ants, etc.,?) clearly showing the difference between mammal-made and invertebrate-made. Further, all the mammal-made burrows uncovered to date in this Upper Middle Jurassic deposit are small to very small, suggesting ancient and very small species adapted to some kind of fossorial or semi-fossorial lifestyles.

While the demarcation line between larger and smaller burrowers is still blurred to say the least, the very small to tiny burrowers are, or appear to be a separate sub-group: prey species and only prey species (since they could not have preyed on anything but insects). Further, we now know – from the Hadrocodium fossil, the Madagascar Trebosphenidians, and the diameter of burrows in both the Navajo and Moab-Entrada - that some of these small critters never increased in size, nor departed from their fossorial lifestyle, during 40 million years, from the very Early to the upper Middle Jurassic. By the Late Jurassic they may have changed their means of survival but the current track, fossil, and burrows record neither support nor reject that possibility.

However It would be hard to believe that these very small to tiny prey species could have possibly altered their methods of survival in the face of constant predation, external or internal. Such alterations would have had to include a series of anatomical and psychological changes, some minor, some substantial. Among them, tree-climbing abilities (switch from burrowing to arboreal lifestyles), abandonment of colony-like safety leading to sharp decreases in population, adaptations to new diets forced by changing habitats, etc., but the biggest alteration of all would have been the 'up-grading' of their status from the lowest of vertebrate prey species to a 'higher position' on the food chain. But no species on Earth can 'vacate' their position on this pitiless food chain without being replaced by another species. Since they were the smallest vertebrates then and still today who could have possibly replaced them at the very bottom of the vertebrate 'Totem Pole'?

No one replaced them. The Morganucodonts are present in the fossil record in exactly the same size, fossorial lifestyle, and the smallest of the vertebrate prey species, from the beginning of the Jurassic to the end of the Cretaceous, a period spanning a staggering 145 million years. A larger period in fact since they are not only present in the Late Triassic but were the ones who crossed the 65 mya K-T Boundary according to conventional thinking. In other words, the Morganucodonts were already 'living fossils' after 145 million years of non-evolution, retaining their highly successful prey species niche from beginning to end. But end of the Cretaceous or today?

That's the question we need to answer. A question that challenges the conventional idea that the mice-like Morganucodonts are the only ancestors of the modern Mammal Class.

The fossil record indicates that Morganucodonts were present in this North-American Sand-Sea. They, or similar species, are most likely if not certainly the ones who made and inhabited the smaller burrows uncovered in the Navajo Sandstone, and in the Moab-Entrada of the upper Middle Jurassic. Some of these burrows are located among ones made by larger mammals, but some are concentrated, and by themselves, in colony-like burrows. Regardless of the situation, all these burrows are inter-connected suggesting that all these earlier mammals were compatible in evolutionary terms. With larger and smaller species playing a different role in the overall survival and well-being of the Early Mammal Class.

Theories to be valid must make predictions that can checked. I predict that burrows made by these Morganucodonts or similar prey species, as partially documented in the Navajo Sandstone and Moab-Entrada, will be uncovered right up to the Late Cretaceous K-T Extinction - and of course, beyond (24).

Not if, but when they do, these burrows will suggest the following:

1. Early mammals survived the K-T Extinction because of their numbers. The survival rate among species is largely if not solely based upon their numbers at the time of the extinction or whatever sporadic natural causes (epidemics, climate and habitat changes, etc.). From a ratio standpoint - regardless of the methods of survival - dinosaurs could not have competed with these multitudes of mammals - and didn't.

2. The smallest of the Jurassic-Cretaceous mammals were per force prey species, thus the most prolific of the Class (as they are still today). Due to their numbers and size they must have been the largest group of early mammals that managed to cross the K-T Boundary.

3. The so-called 'Explosion of Mammals' following the K-T Mass Extinction, presumably due to the demise of predatory dinosaurs, needs to be reviewed. Mammals had already 'exploded' by the Early Jurassic - if not before.

4. Fossorial or semi-fossorial lifestyles were, along with numbers, the main perhaps only reason for mammals to survive the massive K-T Extinction, a projection already made by many paleontologists. However, larger predatory species like Repenomamus giganticus may or may not have been fossorial, at least to some extent. So, unless larger burrows matching their size are uncovered, they may not have been able to cross the K-T Boundary.

5. Unless falsified by discoveries of only small to very small burrows in the mid-late Cretaceous, the two groups that inhabited the Navajo burrows were the main branch of the mammal lineage. The larger types later evolving into varieties of terrestrial and aquatic animals, and the small to tiny types retaining their highly successful fossorial lifestyles from the Jurassic to modern times, but only as barely evolved prey species.

Since these types of burrows are very common in modern times they indicate a continuity in this effective method of survival, and a continuity in the anatomical and psychological aspects of small prey species. This appears to be the case among many contemporary rodents. Placental reproduction could only have enhanced their chances of survival not transforming them overnight into wide-ranging surface predators nor large herbivorous species. So, within this context, contemporary rodents that have retained this form of survival along with anatomical features similar or very close to their Jurassic ancestors, can reasonably be referred to, as 'living fossils'.

Reasonably also because many contemporary animals are undeniable living fossils such as crocodiles, turtles, termites, sharks, scorpions, lungfishes, crayfishes, to name a few. To deny that the Mammal Class was exempt to such possibility would be arbitrary and inconsistent with current discoveries. One of them, unknown to most US paleontologists, took place in 1982, and was published in 1983 in the December issue of the US science magazine Discover.

The superbly preserved fossil of a rat-like animal was uncovered in the now famous Grube Messel pit near Frankfurt, Germany. Dated at 50 million years – the Eocene - It was uncovered along with also superbly preserved fossils of early tapirs, horses, and crocodile species. Documentation was made by the Frankfurt Senckenberg Institute under the direction of paleontologist Dr. Jens Franzen.

Interestingly this rat-like fossil was discovered more than 20 years ago – way before Hadrocodium and Eomaia Scansoria – but never received the attention it deserved, not even a proper scientific name standard for such fossil. In this prolific pit it was only a 'rat' among vastly more interesting Eocene species. Again a reflection of the lack of interest in early mammals and their general unimportance. Its classification as an 'insectivore' is also on par with the conventional view.

However its presence among true mammals, already close to very close anatomically to contemporary species, strongly suggests that small rat-like prey species had not only crossed the K-T Boundary but were firmly established 15 million years after that Mass Extinction. Whether or not that particular creature became extinct after the Eocene Period it's more than reasonable to assume that anatomically similar and prolific prey species not only did not become extinct but survived until contemporary times. Further yet, by that time – 50 mya – it seems unlikely that such small prey species had either 'died-out' or evolved into other types of mammals. The later would assume that such small rat-like species were able in such a short (geologic) time to evolve into other forms of mammals then regress back to their original anatomy, sizes, and methods of survival. Regressive evolution is still no part of the scientific world.

Regressive evolution is not the same as adaptive evolution or de-evolution, meaning the same thing. The terrestrial rodent-like prey species could only have evolved into somewhat similar prey species solely adapted to terrestrial lifestyles and methods of survival. The sole exemption to that rule are the early mammals that inhabited the Ocean shores. Fossorial or not, at least some of these terrestrial but 'maritime' species used the Ocean as their primary source of food. Some eventually became semi-aquatic, like otters and seals, others further de-evoluated into fully aquatic mammals, like whales. Otters and seals are anatomically similar, but in whales the de-evolution of hind limbs into a form of propulsion (tail) and the modification of front limbs into paddles and directional features made them fully aquatic. When this transition between terrestrial to semi-aquatic or aquatic lifestyles took place is still unknown. But it is safe to assume that this transition must have taken place sometime before the K-T Mass Extinction, especially for the fully aquatic whale family. In other words, the only mean for these 'maritime' species to survive the 65 million years K-T Mass Extinction was the Ocean, a water environment they had already adapted to under one form or another, and this for millenniums.

For the fully terrestrial rodent-like prey-species it was a different 'ball-game'. No de-evolution into anything but what they already were in the Mesozoic. The Eocene Grube Messel rat-like fossil is enough by itself to suggest that no 'de-evolution' took place in the ranks of the rodent-like prey species after the 50 million years mark. If (or when?) similar fossils, including tracks and burrows, are uncovered in the 15 million years gap between the 65 mya K-T Mass Extinction and this 50 million years fossil, it would confirm beyond any doubts that at least some small contemporary rodent prey species are indeed 'living fossils'.

Chapter Index

1. Based on my, and other field researchers experience, large numbers of paleontologists still brush tracks aside as unimportant to the study of prehistoric animals. Following the discovery of the outstanding 'Squirrel site' in 2001 Francis 'Fran' Barnes and I reported it to three regional paleontologists, one in a very high position. No answers were ever received. Upon pursuing the matter by telephone I was informed by one of them (whom by courtesy I will not name) that, " I was wasting my time....There is nothing to learn from tracks ".

2. From the late Francis 'Fran' Barnes and his paleobotanical associates.

3. First uncovered in-around wet or swampy horizons, then later scattered here and there in the slickrock of the less eroded segments of the Moab-Entrada. Their length and diameter suggest thin brush around 3-4 foot high.

4. Playas were and still are the most investigated areas of the Navajo since their wet and muddy periphery are the most likely place to find any kind of tracks. To date, only tridactyl tracks (Theropods or Ornithopods?) and tracks of invertebrates have been uncovered in them. The lack of mammal tracks in-around these playas is one of the most puzzling aspects of the Navajo Sandstone. Even more puzzling is the lack of burrows around these fresh water lakes, albeit ephemeral. The only explanation for this total avoidance of fresh water in such an arid environment, seems to be some kind of toxic content, at least to mammals. Puzzling also is the lack of mammal tracks in-around springs. In arid regions a handful of modern rodents have an intestinal system mainly adapted to absorb water via vegetation, although they can lick water from rocks and occasionally drink from fresh water ponds without side-effects. This adaptation doesn't seem possible however for the variety and numbers of primitive mammals that inhabited that region of the Early Jurassic.

5. First developed in the early 1950s by zoologists and botanists in South Africa, then later adapted by range managers of the Bureau of Land Management (BLM) and the US Forest Service to evaluate the effects of commercial grazing on Public Land. For anyone interested in such elaborate studies made (and still up-graded) by combinations of related scientists, they are available at various US Government Offices (at a price). In Africa currently, these still on-going studies are the foundation for the expansion or the sustaining of older Preserves and Parks, including the culling of certain species.

6. An excellent subject for post-graduate or doctoral studies. Nowadays the younger generation is vastly more familiar and accomplished with computers than the older ones, who many a time shun them for their studies.

7. Herbivorous means plant eaters - and only plant eaters. Carnivorous means flesh eaters - and only flesh eaters. But the term omnivorous meaning both flesh and plant eaters is seldom used in paleontological literature when dealing with mammal species. The reason for this ambiguity is mostly due to the lack of comparative studies between ancient and contemporary rodent species. This lack of inter-discipline studies is not restricted to paleontology or geology but is still afflicting many scientific fields, particularly in America. The term insectivore doesn't mean anything unless the dentition of the animal is strictly adapted to insect eating like golden moles, tenrecs, elephant

shrews, etc. including the newly uncovered Late Jurassic Fruitafosssor with teeth solely adapted to termites-ants eating. All the others, ancient and modern, are omnivorous. In Mesozoic times for any one to assert that all early mammals were insectivores not only stretch the imagination but is today scientifically unacceptable.

8. Practically all dinosaur tracks documented to date in the Early and Middle Jurassic of southern Utah are tridactyl (and associated with predators -Theropods of various sizes). Only a handful are currently attributed to herbivorous prey species. The fossil record is somewhat better but still too sketchy to paint a solid picture of these herbivorous species. For the ones interested in this menagerie I highly recommend, 'Dinosaurs of Utah', 1998, by Prof. Frank DeCourten.

9. Prosauropods and true Sauropods had either long or longer necks somewhat or solely adapted to tree or high brush browsing. During the Early Jurassic period (anywhere) this adaptation is confirmed by the fossil record (Massospondylus for instance, a relatively well-known Prosauropod, and Bellusaurus, a small but true Sauropod). To this we add the recent discovery of Lufengosaurus - a superbly preserved and huge (for its time) Prosauropod of Chinese origin dated at the onset of the Jurassic (205-200 mya) whose immense long neck was solely adapted to tree browsing. Whether or not Lufengosaurus (or similar species) were present in southeastern Utah at that time is currently unknown. But what we can at least derive from the fossil and track record is that some kind of high vegetation (trees or high brush) was indeed present during those times - an earlier speculation now confirmed by the recent discovery in the Navajo Sandstone of large trees and numbers of roots attributed to brush-like vegetation.

10. In that context comparative studies mean estimates of 'sustenance' (foraging or flesh) for given dinosaur species. Estimates, or at least educated guesstimates, exist today for most contemporary large species of interest to humans (terrestrial, domesticated or wild). In zoology these studies are called comparative because the knowledge obtained from the study of one given species can be 'transferred' to similar but harder to study animals. Nowadays the use of computerized models permits 'transfers' with much narrower 'latitudes' (margins of error) than was possible only a few years ago.

11. Tridactyl tracks solely and arbitrarily attributed to predatory Theropods were first challenged by Frank DeCourten ('Dinosaurs of Utah', 1998), then later by colleagues and I. At the famed 'Megatracksite' in the Moab-Entrada (upper Middle Jurassic in the Moab, Utah, region), thousands if not millions of tridactyl tracks were solely attributed to migrating Theropods – a blatant inconsistency since Ornithopods were present (via body fossils) in the next upper deposits. Recent re-assessment of the Moab-Entrada deposit, forced by the discovery of mammal burrows, vegetation, swamps, fresh water, bird-like tracks, de-watering pipes, even Ornithischian tracks, etc., has falsified any further ideas that these thousands-millions of tridactyl tracks could have solely been made by Theropods - more so by migrating ones, since none of these animals, Ornithopods included, were migrating anywhere, instead were inhabiting that newly uncovered fairly lush and hospitable deposit.

12. The only exceptions to that rule are continents or islands (Australia, New Zealand, as examples) that became isolated during the splitting of the Pangaea Continent. In such locations certain vertebrate species (like kangaroo or kiwi) managed to survive without specific predators within

or outside their own class. However, when such predators were later introduced by explorers or settlers most of these species quickly became extinct.

13. Bears, particularly grizzly bears, can sometime dig for long periods of time to catch small ground squirrels. As a rule, however, they only dig for them above timberline because, due to the rocky soil, these rodents can only burrow a very short distance. Even for these immensely powerful animals it would be impossible for them to dig for colony-like mammals (like prairie dogs) because their burrows are deep, inter-connected, sometime spread over very large surfaces. For anyone interested in comparative studies of ancient and contemporary predator-diggers I recommend a fascinating account of these animals written by Jack Boudreau, a friend of mine: "Grizzly Bear Mountain", 2000, Caitlin Press Inc. Prince George, BC, Canada.

14. Now confirmed by the Hadrocodium fossil, and this at the very onset of the Jurassic.

15. The 'living proof' of these small, agile, and deadly Theropods, is the cartoonish Arizona Roadrunner, a (mainly) ground bird that feeds on insects, lizards, snakes, birds, and rodents. Whether the Roadrunner is a direct or not descendant of the Theropods it looks and acts like the small ones that inhabited the Jurassic. Its speed and agility are legendary. It many ways the Roadrunner is a modern-day bird-like Theropod that, through the millenniums, had adapted to harsh desert environment and has remained in an ecological niche in which its survival is secure.

16. These 'bird-tracks' were rare to extremely rare until recently. They are now surfacing in increasing numbers as field research is shifting (in the Moab region) from dinosaurs to mammal tracks and burrows. In the Moab-Entrada deposit some are located in-around burrows or within burrow-disturbed areas (Class III burrows). These small to very small tridactyl tracks can be attributed to infant Theropods, very small species of Theropods, or some kind of early birds or 'ground birds'. Whatever the case all of them must have been predatory (early birds included), most likely preying on very small burrowers and insects.

17. The scarcity of comparative studies (teeth-jaws-anatomy-habitats, etc.) between ancient and contemporary mammals is mainly due to the reluctance of most paleontologists to cross into other disciplines.

18. Contrary to their reputation as one of the most vegetarian species on Earth, hippos can and do eat dead flesh, albeit rarely.

19. E. Garcia 'Moles', 1997 - P. Rezende, 'Tracking & the Art of Seeing', 1999. Related to the subject are the lifelong studies of squirrels by Vagn Flyger, a professor (emeritus) of wildlife biology at the University of Maryland.

20. The superbly preserved skeletons of Hadrocodium and Eomaia are undeniable confirmation of such anatomical similarities. In fact so similar that, should some of these early mammals had survived into present times, it would be hard from a distance to distinguish them from some contemporary species.

21. 'Vertebrate Burrows', part of 'The Study of Trace Fossils', Springer-Verlag Publications,

1975. Under the Order Insectivora : "The insectivores as a whole are the most diversified group of mammalian burrowers. At least some members of all the living families are fossorial, and the Talpidae and Chrysochloridae include some of the most highly specialized of all subterranean animals. Shrews (Family Soricidae) are not so highly specialized for digging as moles, but according to Matthews (1971, p. 59), all the approximately 300 species dig burrows. Thanks to Eisenberg and Gould (1970) on the life histories of tenrecs in Madagascar, we now know that most members of this family of 'living fossils' are also burrowers. One genus, Oryzorictes, is very mole-like in adaptations. Note: If this vast family of moles, shrews, etc., are 'living fossils', so are similar-sized rodents with similar fossorial lifestyles.

22. This discovery is barely known in paleontological circles most likely due to the general lack of interest in early mammals. It was pointed out to me by Richard 'Rich' Cifelli, a specialist in early mammals dentition at the Oklahoma Museum of Natural History.

23. Small to very small insect burrows are present in-around many mammal burrows uncovered in the Navajo Sandstone. Some are vertical but most are horizontal. The vertical ones, mostly single tunnels, could have been made by a number of insects ranging from wasps to spiders. The horizontal tunnels appear to have been constructed by either ants or derived termites (Termitinae). All of them have yet to be documented by specialists. However, their importance does not rest in their future determination but in their exploitation by various skeptics and conventionals to quickly reject mammal burrows. 'Termites' being the preferred ones, with 'Crayfish' a close second. Interestingly, very few if any of these skeptics can describe these 'termites' or 'crayfish' burrows. They cannot for the most, since the documentation of these insects and crayfish-lungfish burrows is relatively recent, and generally outside their field. The first comprehensive study of crayfish burrows, made by Steve Hasiotis, was published in 1993 under "Ichnology of Triassic and Holocene cambarid crayfish of North America". The first and only comprehensive study of insect burrows was also made by Steve Hasiotis (U. of Kansas, Dept of Geology). Its title, "Complex ichnofossils of solitary and social soil organisms", published by Elsevier Sciences in 2002. I strongly suggest these 'skeptics' first get acquainted with these studies before blabbering about 'termites' and 'crayfishes'.

24. They already exist, albeit in minimal numbers. Outside the Late Cretaceous burrows recently uncovered at Egg Mountain (Montana - 2002) and partially documented by Anthony Martin (Emory University, Atlanta, Georgia), mammal burrows were discovered and documented in 2004 in the Salt Wash Member of the Upper Jurassic Morrison Formation near the Henry Mountains only 100 miles of so west of Moab, Utah. The documentation was made by Steve Hasiotis, Robert Wellner, Anthony Martin and Timothy Demko, and published in Ichnos, 2004.

Note: Spurred by the recent discovery of immense mammal burrows in southeastern Utah, especially the Class III types, field searches for them are certain to increase in the immediate future. And as soon as field researchers get acquainted with their various configurations I predict they will be uncovered elsewhere throughout the entire Jurassic-Cretaceous period - and beyond.

Recapitulation

The conventional version of the Jurassic-Cretaceous mammal lineage up to 2005:

Three sub-order of Eucynodonts: Tritheledontids, Tritylodontids, Morganucodonts. The Tritheledontids, an extremely mammal-like family, are unknown in the fossil record beyond the Early Jurassic (Navajo Sandstone in Utah). The Tritylodontids, also an extremely mammal-like family, are unknown in the fossil record beyond the Early Cretaceous. The Morganucodonts, a shrew-like (mice-like) family and similar animals, survived until the end of the Cretaceous, and according to the fossil record, are the sole ancestors of the modern Mammal Class.

Tritylodontids were herbivores and apparently abundant in the Early Jurassic, but so far only in what is today China. The Tritheledontids were insectivore and relatively scarce. The Morganucodonts were insectivores and the rarest of the three sub-orders.

The Tritheledontids and Tritylodontids are not part of the Mammal Class, and are classified in receding order of evolution as follows: Therapsids, Synapsids, and Mammal-like reptiles, in accordance with personal views or knowledge of these animals. Only the Morganucodonts are classified as Mammals. Hereby, the conventional view of the mammal lineage in the Mesozoic as **"rare, nocturnal, and insect eaters"**.

In **2005**, two new mammal species were added to the fossil record, both contradicting the Morganucodonts mammal supremacy and the "rare, nocturnal, and insect eaters' conventional view of the mammal lineage in the Mesozoic. Repenomamus giganticus, and its smaller cousin, robustus. The first the size of a small dog, the second smaller, but both carnivorous mammals from the Early Cretaceous. Fruitafossor windscheffeli was a true insectivore from the Late Jurassic, with teeth solely adapted to termites-ants eating. Its dentition challenging the conventional view that Morganucodonts and Tritheledontids were solely insect eaters, a proposition based on their teeth and jaws configuration. At per January 2006, Repenomamus and Fruitafossor have yet to receive a proper classification outside carnivorous and termites-ants eater, but both are true mammals unknown until then in the fossil record.

All the above solely based on the fossil record.

In **2004**, a new dimension was added to the fossil record via the discovery and partial documentation of gigantic burrows and increasing numbers of tracks attributed to mammals or mammal-like animals. The conventional view of the mammal lineage, again solely based on the fossil record, was at once applied to these discoveries. The increasing number of tracks must have been made, at least most of them, by Therapsids, Synapsids, or Mammal-like reptiles, again depending on personal view of these animals. As for the burrows, they must have been made by invertebrates, roots, reptiles, or Tritylodontids since they were apparently abundant during that period, at least in China. Conventional views designed to shore-up the now shaky proposition that Morganucodonts were the sole ancestors of the modern Mammal Class.

As per January **2006**, these Navajo burrows indicate, via the diameter of the tunnels, a variety of animals from mouse-like to badgers. Hereby, the smaller tunnels could have been made by

Morganucodonts or similar species, the larger ones by any kind of mammals from Tritylodontids to other species still unknown in the fossil record. Until their skeletons are uncovered to project them as Therapsids, Synapsids, or Mammal-like reptiles, is both arbitrary and inconsistent with the biological nature of colonial animals. Whoever these animals were, they lived together in interconnected city-like burrows. Biology by itself nullify propositions of communal living between less and more advanced species, more so when differences in size strongly suggest the presence of small prey species within these colony-like burrows.

Therefore, the animals who made and inhabited these gigantic burrows must have been compatible to each other in evolutionary terms, whether they were predators, prey species, or enjoyed communal living regardless of their size and means of survival. And the only vertebrates in the Jurassic and the Cretaceous who could have burrowed and lived together in such manner were mammals, or very close to be. This biological or symbiotic relationship indicates that, regardless of their past or future classification, all these animals belonged to the Mammal Class in one form or another. And that the conventional view of the mice-like Morganucodonts as the sole ancestors of the modern Mammal Class, including past classifications of similar species as 'reptilian', is in great need of revision.

Looking back to looking forward

By the time, or sometime after you finish reading this book, some of its speculations, projections, or conclusions, will be enhanced or altered by new discoveries, new interpretations, or studies, unknown until then. Science is an ever evolving process, never standing still, especially when faced with complex issues. Even with these realities factored in, the general context of this book should remain within the boundaries of scientific acceptance.

These burrows, vegetation, de-watering pipes, and other physical 'proxies' are not about to disappear. Instead, they are going to expand both in numbers and scientific understanding. Some are already protected by National Parks boundaries, others will fall in the immediate future, under Federal and State Protection. Precious scientific material available to all.

Looking back, these discoveries have been a struggle between field work and academic acceptance. Field work hampered by lack of interest, financial and physical constraints, but also by lack of competence in related disciplines among some field researchers and academics alike. It took eight long years for these burrows and related discoveries to barely surface into academic and professional circles. Long years indeed for the ones whose sole support was their dedication to scientific advances.

Looking forward, the struggle is far from over but has entered a new arena where these discoveries can no longer be dismissed as allegations, or impossible under conventional wisdom. Scrutiny has now replaced skepticism. But in the scientific world, scrutiny is a very slow process, with long years ahead of us before a consensus emerges among various academic circles. What this consensus will be is too far into the future to guess. At this early stage, the only thing we can project is a healthy, and most likely contentious debate about the real profile of our Mesozoic ancestors:

The objective of this book.

Acknowledgements

While the accompanying bibliography contains the main scientific background for this book others did contribute in lesser ways but nevertheless were important to its deductions. To these individuals who will recognize their input throughout these pages I extent my sincere thanks for their benevolent and impartial contributions.

Above and beyond these contributions I wish to recognize certain scientists to whom we are indebted, and who still contribute to the new understanding of the mammal presence in the Jurassic, particularly the early part, at a time when such deductions and projections were not part of the scientific mainstream.

M.R. Voorhies. In his superbly documented manuscript, 'Vertebrate Burrows', published in 1975 by Springer-Verlag, New-York-Berlin, under the title, The Study of Trace Fossils, M.R. Voorhies laid the zoological and morphological foundation for these then unknown Jurassic burrows.

Spencer Lucas. The first paleontologist to recognize these gigantic Navajo burrows as most likely made by some kind of early mammals, hereby starting a chain of investigations still going-on today. His initial courage in the face of peers' skepticism if not outright rejection set him apart in his own discipline.

Tamsin McCormick. Her skeptical attitude and attention to details greatly helped in tightening the assessment of each discovery, particularly vegetation and trees. This skepticism was and still is an extremely valuable shield against premature interpretations when dealing with discoveries unknown or barely known in scientific circles.

Colin Egan. Along with Spencer Lucas, the first professional to recognize these burrows as mammal-made, but from a biological perspective. His contributions are noteworthy, as he is still one of the few scientists challenging determinations solely based on one discipline.

Donald Rasmussen. His knowledge of water-tables, aquifers, de-watering pipes, and related wet horizons have permitted to expand our knowledge of the Navajo Sandstone way beyond the discovery of these burrows.

Steve Hasiotis. Apart from his expertise in vertebrate and invertebrate burrows his advocacy of 'proxies' to determine and interpret complex issues is noteworthy since this methodology is still considered avant-garde in many scientific circles.

Lastly, I would like to recognize my best friend John Andre, and his wife Joan. Without their support I would never have been able to put together this and other books and manuscripts I had written earlier.

References

Ahlbrandt, T. & Andrews, S. 1978. Bioturbation in aeolian deposits. Article. Journal of Sedimentology Petrology

Airoldi, J. 1976. Le comportement fouisseur du campagnal terrestre, Arvicola terrestris sherman (Mammalia, Rodentia). Article. Revue Suisse, Zoologie

Anderson, D. 1988. Tunnel construction methods and foraging path of a fossorial herbivore, Geomys bursatius. Article. Journal of Mammalogy

Aulak, W. 1967. Estimation of small mammal density in three forest biotopes. Article, Ekologia Polska, Warsaw, Poland

Baars, D. 1983. The Colorado Plateau – A Geologic History. University of New Mexico Press, Albuquerque, NM

Barnes, F. 1998. Navajo Sandstone, A Canyon Country Enigma. Canyon Country Publications, Moab. Utah

Barnes, F. 2000. Canyon Country Geology. Canyon Country Publications, Moab, Utah

Barnes, F. 2003. Footprints on the shores of time. Canyon Country Publications, Moab, Utah.

Benton, M. 1988. Burrowing in vertebrates. Article. Nature

Blakey, R. 2005. Paleography of the Navajo Sandstone, Colorado Plateau and Vicinity. Article. Journal of the Dan O'Laurie Museum, Moab, Utah

Brown, L. 1972. Unique features of the tunnel system of the eastern mole in Florida. Article, Mammal Journal

DeCourten, F. 1998. Dinosaurs of Utah, University Press, Salt Lake City

Dinesman, L. 1971. Mammalian lairs in paleoecological studies and palynology. Article, Palynol Journal

Eisenberg, J. & Gould, E. 1972. The tenrecs: a study in mammalian behavior and evolution. Paper. Smithsonian contribution, Zoology. 27: 1-137

Ellenberger, P. 1975. L'explosion demographique des petits quadrupedes a l'allure de mammiferes dans le Stromberg Superieur (Triassique) d'Afrique du Sud apercu sur les origines du Permien. Article (and Presentation). Centre National de la Recherche Scientifique, Paris, France

Falkowski, P. & all. "Did rising oxygen levels fuel mammal evolution?". Article in Science Journal & National Geographic News, published Sept. 30, 2005

Faurie, A., Dempster, E. & Perrin, M. 1996. Foot drumming patterns of southern African elephant shrews. Article. Mammalia

Flake, L. 1973. Food habits of four species of rodents on a short-grass prairie in Colorado. Article. Journal of Mammalogy

Foster, J. 2005. The Navajo Sandstone and Its Fossil Vertebrates: More to come?. Article. Journal of the Dan O'Laurie Museum, Moab, Utah

Gleannie, K. & Evamy, B. 1968. Dikaka: plants and plant-roots structures associated with aeolian sand. Article. Paleogeography, Paleoclimatology, and Paleoecology. 4, 77-87

Gore, R. 2003. The Rise of the Mammals. Article, National Geographic Magazine, April issue.

Groenewald, G, Welman, J., & MacEachern, J. 2001. Vertebrate Burrow Complexes from the Early Triassic Cynognathus Zone (Driekoppen Formation, Beaufort Group) of the Karoo Basin, South Africa. Article. Society for Sedimentary Geology

Hansen, R. 1965. Pocket gopher density in an enclosure of native habitat. Article. Journal of Mammalogy

Hasiotis, S. 1993. Ichnology of Triassic and Holocene cambarid crayfish of North America: an overview of burrowing behavior and morphology as reflected by their burrow morphologies in the geological record. Paper. Dept. of Geological Sciences, University of Colorado, Boulder, Colorado

Hasiotis, S. & Mitchell, C. 1993. A comparison of crayfish burrow morphologies: Triassic and Holocene fossil, paleo-neo-ichnological evidence, and the identification of their burrowing signatures. Article. Ichnos

Hasiotis, S., Wellner, R. & Dubiel, R. 1993. Application of morphologic burrow interpretations to discern continental burrow architects: lungfish or crayfish. Article. Ichnos

Hasiotis, S. 2002. Complex ichnofossils of solitary and social organisms: understanding their evolution and roles in terrestrial paleoecosystems. Article. Elsevier Science

Hasiotis, S., Wellner, R., Martin, A. & Demko, T. 2004. Article. Vertebrate burrows from Triassic and Jurassic continental deposits of North America and Antartica: their paleoenvironmental and paleoecological significance. Ichnos

Haubold, H., & Lucas, S. 2001. Tetrapod footprints of the Lower Permian Choza Formation at Castle Peak, Texas. Article. Palaontologishe Zeitschrift, 2003, Institut fur Geologishe Wissenschaffen, Halle, Germany

Heckert, A. & Lucas, S., Editors. Vertebrate Paleontology in Arizona. Bulletin 29, New Mexico Museum of Natural History & Science, Albuquerque, NM

Heth, G. 1991. Evidence of above ground predation and age determination of the preyed, in subterranean mole rats (Spalax ehrenbergi) in Israel. Article. Mamalia 55-529-542

Hibbard, C. 1967. New rodents from the Late Cenozoic of Kansas. Paper. Michigan Academy of Science, Arts, and Letters

Hildebrand, M. 1985. Digging in quadrupeds. Article part of Functional Vertebrate Morphology. Belknap Press of Harvard University Press, Cambridge, Massachusetts

Jarvis, J & Sale, B. 1970. Burrowing and burrow patterns of East African mole-rats Tachyoryctes, Heliophobius and Heterocephalus. Paper. Zoology Dept. University of East Africa, Nairobi, Kenya

Jarvis, J., Bennett, N., & Spinks, A. 1998. Food availability and foraging by wild colonies of Damaraland mole-rats (Cryptomys damarensis): implications for sociality. Article. Oekologia 113, 290-298

Jenkins, F., Crompton, F, & Downs, W. 1983. Mesozoic mammals from Arizona: new evidence on mammalian evolution. Article, Science (v. 222)

Kocurek, G. 2005. Navajo Sandstone – The Record of a Great Sand Sea in earth History. Article. Journal of the Dan O'Laurie Museum, Moab, Utah

Lessa, E. & Thaeler, C. 1989. A reassessment of morphological specializations for digging in pocket gophers. Article. Journal of Mammalogy

Lockley, M. & Hunt, A. 1995. Dinosaur Tracks and other fossil footprints of the western United States. Columbia University Press, New York

Lockley, M. 2002. Fossil Footprints of the World. Lockley-Petersen Publication. Denver, Colorado

Lockley, M. & Foster, J. 2003. Late Cretaceous Mammal Tracks from North America. Article. Ichnos 10: 269-276

Lockley, M., Lucas, S., Hunt, A., & Gaston, R. 2004. Ichnofaunas from the Triassic-Jurassic Boundary Sequences of the Gateway area, Western Colorado: Implications for Faunal Composition and Correlations with Other Areas. Article.Ichnos

Lockley, M. 2005. Enigmatic Dune Walkers from the Abyss: Some thoughts on water and Track Preservation in Ancient and Modern Deserts. Article. Journal of the Dan O'Laurie Museum, Moab, Utah

Lockley, M, Odier, G, Mitchell, L. & Mickelson, D. Small Theropod tracks assemblages from Upper Jurassic eolianites (Moab Member of the Curtis Formation) of eastern Utah: insights into Jurassic Ecosystem. Documentation for BLM Regional Office, Moab, Utah.

Loope, D., Rowe, C., & Joeckel, R. 2001. Annual monsoon rains recorded by Jurassic dunes. Article. Nature

Loope, D. & Rowe, C. 2003. Long Lived Pluvial Episodes during Deposition of the Navajo Sandstone. GSA, Journal of Geology

Lucas, S. & Morales, M. 1993. The Nonmarine Triassic. New Mexico Museum of Nartural History and Science, Bulletin 3

Lucas, S. & Heckert, A. 1995. Early Permian Footprints and facies. New Mexico Museum of Natural History and Science, Bulletin 6
Lucas, M. & Morales, M. 1993

Lucas, S., Guex, J., Tanner, L., Taylor, D., Kuerschner, W., Atudorei, V., & Bartolini, A. 2005. Definition of the Triassic-Jurassic Boundary. Albertiana 32

McCrea, R. & Sarjeant, W. 2001. New ichnotaxa of bird and mammal footprints from the Lower Cretaceous, (Albian) Gates Formation of Alberta. Article. Mesozoic Vertebrate Life, Indiana Press University, Bloomington, Indiana

McNab, B. 1966. The metabolism of fossorial rodents: a study of convergence. Article, Ecology

Miao, D. 1988. Skull morphology of Lambdopsalis bulla (Mammalia, Multituberculata) and its implications to mammalian evolution. Contributions to Geology, University of Wyoming, special paper 4, viii and 104pp

Miller, M., Hasiotis, S., Babcock, L., Isbell, J., & Collinson, J. Tetrapod and Large Burrows of Uncertain Origin in Triassic High Paleolatitude Floodplains deposits, Antarctica. Palaios, 2001, p.218-232

Morgan, G. & Lucas, S. 2000. Pliocene and Pleistocene vertebrate faunas from the Albuquerque Basin, New Mexico. Bulletin 16. New Mexico Museum of Natural History and Science.

Odier, G. 2003. The Jurassic: A New Beginning?. Trafford Publishing, Victoria, BC, Canada

Olsen, P & Galton, P. 1977. Triassic-Jurassic extinctions: are they real?. Article, Science (197:983-86)

Olsen, P. & Padian, K. 1986. Earliest record of Batrachopus from the southwestern United States. Paper in, The beginning of the age of dinosaurs, Cambridge University Press, New York

Padian, K. 1986. The origin of birds and the evolution of flight. Paper. California Academy of Sciences

Powell, L. 1999. Night Comes to the Cretaceous – Dinosaurs extinction & the transformation of modern geology. W.H. Freeman & Co., New York

Reed, C. 1954. Some fossorial mammals from the Tertiary of western north America. Article. Paleontology Journal

Reichman, O. & Smith, S. 1985. Impact of pocket gopher burrows on overlying vegetation. Article. Journal of Mammalogy.

Robertson, D., McKenna, M., Toon, O., Hope, S., & Lillegraven, J. 2004. Survival in the first hours of the Cenozoic. GSA Bulletin, May/June 2004

Sarjeant, W. & Thulborn, R. 1986. Probable marsupial footprints from the Cretaceous sediments of British Columbia. Article. Canadian Journal of Earth Sciences

Schultz, J., Lockley, M., & Gaston, R. 1996. First reports of synapsid tracks from the Wingate and Moenave Formations, Colorado Plateau Region. Article. Continental Jurassic Symposium Volume, Museum of Northern Arizona, Bulletin 60. Flagstaff

Sheets, R. 1971. Burrow systems of prairie dogs in South Dakota. Article. Journal of Mammalogy.

Shipman, T. & Parrish, J. 2005. Limestone Within the Navajo Sandstone and Their Climatic Significance. Article. Journal of the Dan O'Laurie Museum, Moab, Utah

Smith, R. 1986. Helical burrow casts of therapsid origin from the Beaufort Group (Permian) of South Africa. Paper. Dept. of Karoo Paleontology, South African Museum, Cape Town, South Africa

Stokes, W. 1978. Animal tracks in the Navajo-Nugget Sandstone. Article. Geology, University of Wyoming.

Stokes, W. & Madsen, J. 1979. Environmental significance of pterosaur tracks in the Navajo Sandstone, Jurassic, Grand County, Utah. Geological Studies. Brigham Young University, Provo, Utah.

Sues, H. 1986. The skull and dentition of two tritylondid from the lower Jurassic of western north America. Bulletin. Harvard University Museum of Comparative Zoology

Verlander, J. 1995. Basin scale stratigraphy of the Navajo Sandstone: southern Utah, USA. Phd thesis. University of Oxford

Voorhies, M. 1975. Vertebrate burrows. Article. The Study of Trace Fossils. Springer-Verlag, New York-Berlin

Wilks, B. 1963. Some aspects of the ecology and population dynamics of the pocket gophers (Geomyidae, Rodentia). Article. American Nature 96, 303-316

Wilson, D. 1975. The adequacy of body size as a niche difference. Article. American Nature 109, 769-784

Winkler, D. 1991. Life is sand sea: Biota from Jurassic interdunes. Paper. Shuler Museum of Paleontology, Dallas, Texas

Yaoming, H., Jin, M., Yuanqing, W., & Chuankui, U. 2005. Large Mesozoic Mammals Fed on Young Dinosaurs. Article. Nature, Jan. 2005

Addenda

Another major and timely discovery

Reported by Randolph E. Schmid of the Associated Press on February 24, 2006, and taken from an article published on the same date by Science Magazine. At that time this book was already in press, thus the following is a direct quote from the AP Press Release.

Under the title, "Fossil alters notions about mammals" this AP Release summarizes the its scientific implications: "The discovery of a furry, beaver-like animal that lived at the time of the dinosaurs has overturned more than a century of scientific thinking about Jurassic mammals". This fossil was uncovered in the Inner Mongolia of China by a team led by Quang Ji of the Geological Sciences in Beijing. It is dated at 164 million years, the beginning of the Late Jurassic Period, also known as the 'Jurassic Park' of the dinosaurs.

Zhe-Xi Luo, the well-kown and respected curator of vertebrate paleontology at the Carnegie Museum of Natural History in Pittsburgh, commented about this discovery in the following manner: " The ecological role of mammals in the time of dinosaurs was far greater than previously thought. The animal is the earliest swimming mammal to have been found and was the most primitive mammal to be preserved with fur, which is important to help keep constant body temperature. For over a century, the stereotype of mammals living in that era has been of tiny, shrew-like creatures scurrying about underbrush to avoid the giant creatures that dominated the planet".

The discovery in 2005 of the two Early Cretaceous Repenomamus (Giganticus and Robustus), carnivorous mammals feeding on infant dinosaurs, followed almost immediately by the Late Jurassic Fruitafossor windscheffeli, a termite eater, had already shaken the pre-conceived idea of early mammals as 'small, nocturnal, and insectivores'. These two major discoveries came to the surface after the 2004 documentation of immense mammal burrows in the Early Jurassic of Utah. This new and compelling Chinese discovery again re-confirms that early mammals were not only a very diverse lot but also were, numerically speaking, the dominant species during that period, albeit the smallest of the vertebrates.

This newly-discovered mammal had a flat, scaly tail like a beaver, vertebra like an otter and teeth like a seal. It swam in lakes and fed on fish. The reason for its name: Castorocauda lutrasimilis, meaning castor from the Latin beaver, cauda for tail, lutra for river otter and similis meaning similar. Thomas Martin of the Research Institute Senckenberg in Frankfurt, Germany, (but not a member of the Chinese team), commented that "the discovery pushes back the mammal conquest of the waters by more than 100 million years". It does more than that: It forces the scientific community to re-assess the origin, radiation, and evolution of mammals from the Early Jurassic onward.

TETRAPOD BURROWS FROM THE LOWER JURASSIC NAVAJO SANDSTONE, SOUTHEASTERN UTAH

SPENCER G. LUCAS[1], KATRINA E. GOBETZ[2], GEORGES P. ODIER[3], TAMSIN MCCORMICK[4] AND COLIN EGAN[5]

[1]New Mexico Museum of Natural History and Science, 1801 Mountain Road NW, Albuquerque, NM 87104; [2]Department of Biology, MSC 7801, James Madison University, Harrisonburg, VA 22801; [3]115 W Kane Creek Blvd, no. 29, Moab, UT 84532; [4]Plateau Restoration, PO Box 1363, Moab, UT 84532; [5]114 W Kane Creek Blvd, no. 28, Moab, UT 84532

Abstract—Northwest of Moab, Utah, the Lower Jurassic Navajo Sandstone contains sedimentary structures we interpret as casts of tetrapod burrows at two localities that extend over several square kilometers. These burrow casts have subcircular cross sections and consist of shallow, low-angle ramps leading to multiple terminal chambers. Branching is irregular and variable, no burrow linings or traces are visible, and burrow walls are smooth. The burrow casts cut across bedding planes and contain a fill that is identical to the host rock, but the casts preferentially weather from the host rock. Most of the burrow casts have a diameter of 10-20 cm, and in many places form complex mazes that are concentrated in elevated mounds of 1-2 m diameter. Locally, the upper part of the burrowed interval also contains rhizoliths, which are readily distinguished by their knobby surface texture, relatively small diameters (< 10 cm), distinct (calcareous or siliceous) infilling, vertical orientation, and downward branching. The tetrapod burrow casts in the Navajo Sandstone most resemble the burrows and elevated nest mounds of modern Mediterranean blind mole-rats. The most likely producers of the Navajo burrows were tritylodontid cynodonts, whose skeletal anatomy indicates that they were fossorial scratch diggers. The extensive burrowed horizons in the Navajo Sandstone suggest that large populations of herbivorous tritylodontids inhabited interdune environments during intervals of high rainfall (pluvials) of the Navajo erg, and the complexity of the burrow mazes indicates some degree of sociality among tritylodontids.

INTRODUCTION

The Lower Jurassic Navajo Sandstone represents a vast sand sea (erg) that covered an area of more than 265,000 km^2 in five states of the American West (e.g., Kocurek and Dott, 1983). Vertebrate ichnology of the Navajo Sandstone has long focused on its tetrapod footprint record (Lockley and Hunt, 1995; Rainforth and Lockley, 1996; Lockley, 2005). Here, we document tetrapod burrow casts in the Navajo Sandstone near Moab, Utah (Fig. 1). Fran Barnes first discovered such burrow casts, and they have been mentioned in print by Odier (2004), Odier et al. (2004), Eisenberg (2005) and Ostapuk (2005), among others, but have yet to be thoroughly described, illustrated, and evaluated. Herein, we provide such thorough documentation of the Navajo Sandstone burrow casts and discuss their paleoecological and paleobiological significance. Throughout this paper, NMMNH refers to the New Mexico Museum of Natural History and Science, Albuquerque.

Extensive burrowed horizons are present at several stratigraphic levels in the Navajo Sandstone northwest of Moab. Here, we focus on two exceptionally widespread and well-preserved burrow-cast localities, both in the lower part of the Navajo Sandstone (Fig. 1). These sites are designated NMMNH localities 5680 and 5681, and each preserves an intensively burrowed interval that covers square kilometers of outcrop (Figs. 2-3).

DISTINCTIVE FEATURES OF VERTEBRATE BURROWS

When Voorhies (1975, p. 332) summarized the fossil record of vertebrate burrows he knew of no examples of Mesozoic tetrapod burrows: "the 'Age of Reptiles' is notable for the absence of any known vertebrate burrows." However, since Voorhies' article, various Mesozoic vertebrate burrows have been described, especially from the Triassic strata of the Karoo basin in South Africa (Smith, 1987; Groenewald, 1991; Groenewald, et al., 2001). Putative burrow casts have also been described from Triassic and Jurassic strata in the American Southwest by Hasiotis et al. (2004), but we are not convinced that all of these structures are burrow casts. Thus, for example, although some of the "large-diameter burrow complexes" in the Upper Jurassic Morrison Formation illustrated by Hasiotis et al. (2004, fig. 6; also see Hasiotis, 2002, p. 122, figs.B-C) appear to us to instead be a mixture of rhizoliths and calcrete nodules.

No widely recognized criteria for the identification of fossil vertebrate burrows have been articulated in the literature. Therefore, Table 1 lists features that we consider characteristic of fossil vertebrate burrows. Not all of these features are present in all vertebrate burrows, and some of the features are not unique to vertebrate burrows. To identify a structure as a vertebrate burrow, we must rule out the other possibilities, which are that it is an invertebrate burrow, plant root cast (rhizolith) or inorganic structure such as a concretion.

Invertebrate Burrows

The basic architecture of invertebrate and vertebrate burrows may converge on similar morphologies, due to paleoenvironmental and paleoclimatic factors and the association between organism and substrate (Hasiotis et al., 2004). Specific behaviors, such as aestivation in lungfish and amphibians (Romer and Olson, 1954; McAllister, 1991; Hembree et al., 2004; 2005) also result in vertebrate burrows that resemble invertebrate traces in their vertical orientation and repetitive morphology. Given these possible similarites, however, invertebrate burrows are generally much smaller in diameter than are vertebrate burrows (Bromley et al., 1975, discuss some rare exceptions to this). Typically, invertebrate burrows are vertical (perpendicular to bedding) or horizontal (parallel to bedding), not inclined to bedding, as are most vertebrate burrows. Invertebrate burrows often have walls that are ornamented or textured due to the packing of burrow walls or meniscate backfilling by the burrow maker (Hasiotis and Mitchell, 1993; Hasiotis, 2002). In contrast, vertebrate burrow casts usually lack this type of surficial morphology, and instead show linings, scratch marks, or occasionally grooves made by teeth (Smith, 1987; Martin and Bennett, 1977; Groenewald et al., 2001; Gobetz, In press). In addition, invertebrate burrows are characteristically of much more regular and repetitive shapes than are vertebrate burrows, which are usually of irregular shape. Mammalian burrows, for instance, usually have vertical entrance shafts leading to a main low-angle or horizontal tunnel with secondary and tertiary branches (Voorhies, 1975; Nevo, 1999). The majority of ichnological research has focused on invertebrate burrows, resulting in precise criteria for recognition of invertebrate burrow types, which help to eliminate possible confusion of invertebrate burrows with vertebrate burrows.

Rhizoliths

Klappa (1980; also see Ekdale et al., 1984) provided a basis for distinguishing between rhizoliths and burrows of all kinds: (1) rhizoliths generally have circular cross sections and are cylindrical or conical, whereas burrows are of more variable shapes; (2) rhizoliths bifurcate downward, and diameters decrease downward, whereas burrows rarely bifurcate downward, and, if they do, diameter remains consistent; (3) rhizoliths form either vertical branching systems or horizontal ramifying networks, whereas burrows can have more variable orientations; (4) rhizoliths sometimes contain partially or wholly petrified remnants of internal structures (e.g., vascular tissue) or carbon films, whereas burrows do not; and (5) rhizoliths have few external features, whereas burrow walls may show striations, scratch marks or other kinds of surficial morphology. Rhizoliths are also smaller (0.1 mm-20 cm diameter) than most vertebrate burrows (Klappa, 1980).

Inorganic Structures

Inorganic concretions are usually formed by diagenetic or post-diagenetic (often groundwater) processes. They typically lack the tubular and relatively smooth-walled shapes characteristic of most vertebrate burrow casts. Furthermore, concretions do not form interconnected networks and otherwise lack the architectural organization that istypical of vertebrate burrows: one or more surface openings, main shaft with secondary and tertiary tunnels branching therefrom, and one or more terminal chambers.

DESCRIPTION OF NAVAJO BURROW CASTS

Architectural Morphology

Odier (2004, p. 53-60) provided the first published description of the tetrapod burrow casts in the Navajo Sandstone northwest of Moab. He described the casts as cylinders with diameters of 5 to 18 cm, and recognized three "classes" among them, based on associated paleoenvironment: (1) class I burrows, which are located on ledges and are characterized by tunnels running at or near the surface and; (2) class II burrows on flat, interdunal surfaces; and (3) class III burrows, which consist of networks inside surface mounds.

Comprehensive morphological description of the Navajo burrow casts is difficult, due to the complexity of branching patterns among individual burrows and the networks or mound-forming labyrinths in which they occur. We describe the architecture of these casts using the terminology of Hickman (1990) and Nevo (1999). Shafts and tunnels, respectively, refer to vertical and horizontal portions of burrow systems, after Hasiotis et al. (2004).

The Navajo Sandstone burrow casts are mostly tubular structures that are sub-circular in cross section and measure 10-19 cm in diameter. Most primary tunnels or chambers have multiple branches, and/or are connected to other burrows (Figs. 2E-F, 3B-D, 4F-G). Some burrows end in cul-de-sacs that are either expanded and rounded bulbs (Fig. 4D-E) or more angular-shaped chambers that give the burrow a boot-like shape (Fig. 4A-C). Chambers of similar shape are known from other fossil vertebrate burrows (Morgan and Lucas, 2000), some of which have traces indicating excavation by a vertebrate (Gobetz, In press). Some of the cul-de-sacs on the Navajo burrows consist of large, irregularly-shaped chambers with multiple chamber entrances (Figs. 2F, 3B-C, 4F-G). The burrow casts are either inclined, horizontal or (rarely) vertical tubes that cut across bedding planes (Figs. 2D, 3D). In general, the burrow casts form complex, labyrinthine networks, some of which comprise elevated mounds, apparently constructed above the paleosurface, of 1-2 m diameter (Fig. 2B-D). These mounds are composed of complex networks of burrows that are typically interconnected and connected to various chambers and cul-de-sacs. Mound diameters are 1-2 m. Some of the mounds are connected to each other by horizontal burrows. No fossil remains have been found in or associated with the Navajo Sandstone burrow casts. Detailed mapping of the Navajo Sandstone burrow systems is needed to establish their exact architectural design, spatial geometry, and density.

Surficial Morphology

Burrow walls are smooth, without inner linings or traces such as longitudinal ridges or grooves. The burrow filling is the same as the surrounding matrix, but the burrows are preferentially cemented, so that they differentially weather out from the matrix (Figs. 2-3).

DISCUSSION

Why Are These Vertebrate Burrow Casts?

The Navajo Sandstone burrow casts described here display 8 of the 11 characteristics listed in Table 1 as typical of vertebrate burrows. Thus, they have distinctive shapes and subcircular to elliptical burrow cross sections, cut across bedding, are preferentially preserved, have one or more secondary branches or entrances into chambers, variable branching and multiple terminal chambers, and shallow vertical shafts that lead to low angle ramps or an underground maze. The Navajo burrows do not have a fill that contrasts with the host rock, but not all fossil vertebrate burrows show this feature. For example, Gee et al. (2003) described a Miocene rodent burrow system that contained a fill with similar color and grain size to the surrounding sediment, so that tunnels and chambers were invisible on outcrop, unless they contained caches of fossilized nuts.

The Navajo burrow casts are generally similar to the cynodont burrow casts described by Groenewald et al. (2001) and are similar to a variety of mammal burrows, both fossil and extant, in the literature. Indeed, most striking is how similar the Navajo burrow casts, networks and mounds are to the burrow systems created by the Mediterranean blind mole rat, *Nannospalax* [= *Spalax*] (Nevo, 1961; Nowak, 1999). The relatively large size, multiple branches, and living chambers of the Navajo burrow casts, among other features (see above), also preclude an invertebrate burrower.

The consistent architectural morphology and structural complexity of the Navajo burrows is compelling evidence for their biogenic origin; they do not resemble any of the inorganic concretionary structures sometimes seen in the Navajo Sandstone. The burrows also can be distinguished from rhizoliths based on the general criteria that distinguish burrows and rhizoliths discussed above. Furthermore, rhizoliths that co-occur with the Navajo burrows (Figs. 3E-F, 4H-I) are readily distinguished by their rough surface texture (Fig. 4H-I), downward branching pattern (Fig. 2E-F) and replacement by carbonate/silica.

Excavator(s)

Without a fossil skeleton or the bones of tetrapods within or associated with the Navajo Sandstone burrow casts, it is impossible to be certain exactly what animal made them. Nevertheless, the size range and morphology of the burrows, their resemblance to the burrows of some extant mammals and the skeletons of Early Jurassic fossil tetrapods provide clues to possible excavator(s).

Based on such considerations, we conclude that the most likely maker of the Navajo Sandstone burrows was a tritylodontid cynodont (Fig. 5). Other therapsids, such as dicynodonts, have been found associated with fossil burrows (Smith, 1987) or are preserved in coiled positions that infer death inside burrows (Hotton, 1987); thus, fossoriality is known to have been a part of therapsid paleobiology.

Winkler et al. (1991) reported the incomplete skeleton of a tritylodontid (probably *Kayentatherium*: see Sues, 1986) from interdunal deposits of the Navajo Sandstone in northeastern Arizona. No functional morphological analysis of tritylodontids has been published, although Kühne (1956) suggested that the most completely known tritylodontid, *Oligokyphus*, was arboreal, and the skulls and dentitions of tritylodontids are readily recognized as those of herbivores. Nevertheless, Winkler et al. (1991, p. 890) described the Navajo Sandstone tritylodontid as a "long-bodied, somewhat sprawling scratch digger, well adapted to digging in the soft substrate on which it lived."

To support this inference, Winkler et al. (1991) drew attention to the short radius and long olecranon process on the ulna of the Navajo tritylodontid. The olecranon process is two-thirds the length of the ulna, a characteristic adaptation of living scratch diggers such as mole rats, armadillos and aardvarks (Hildebrand, 1974, 1985; Coombs, 1983) for exerting greater force against the substrate (Stein, 2000). Indeed, a variety of features of tritylodontids identify them as fossorial scratch diggers (see Hildebrand, 1985, and Stein, 2000, for a description of scratch-digging and Young, 1947, and Kühne, 1956, for a discussion of tritylodontid postcranial anatomy). These include: large and blunt head, short neck, cylindrical body, relatively short limbs in which the proximal segments are within the body contour, creating a streamlined body plan, robust humerus with hypertrophied medial epicondyle, features of the ulna and radius mentioned above, and laterally-compressed claws. These features provide compelling evidence that tritylodontids were fossorial scratch diggers (cf. Gambaryan, 1960; Hildebrand, 1974, 1985). However, the very large, chisel-like anterior teeth, relatively large skull with stout zygomatic arches and large sagittal and nuchal crests, also suggests the possibility of chisel-tooth digging by tritylodontids (Hildebrand, 1974, 1985).

The Navajo *Kayentatherium* skeleton illustrated by Winkler et al. (1991, fig. 4) has a humerus length of ~ 10.3 cm and a minimum body length of 66 cm (actual total body length would have been much greater, as only the trunk is preserved). The skull of *Kayentatherium* described by Sues (1986) is 28 cm long, so this tritylodontid was large enough to make the largest burrow casts present in the Navajo Sandstone near Moab. However, many of the Navajo burrow casts are smaller than the burrows that would have been made by the rather large (for tritylodontids) specimens of *Kayentatherium* described by Sues (1986) and Winkler et al. (1991). This raises the possibility that smaller tritylodontids (known tritylodontid size ranges down to taxa with skull lengths of 4 cm or less: Luo and Wu, 1994) or mammals made some of the burrows.

Most Early Jurassic mammals, however, are morganucodontans and haramiyids that would have been much too small to have excavated any of the Navajo Sandstone burrows (Kielan-Jaworowska et al., 2004) if we assume that burrow diameter closely matches body size, as in modern mammals (Hickman, 1990). The largest known Early Jurassic mammal, *Sinoconodon*, has a maximum skull length of ~ 6 cm (Kielan-Jaworowska et al., 2004), so it seems unlikely that mammals could have made anything other than some of the smallest Navajo burrows. Indeed, *Fruitafossor*, a fossorial mammal recently described from the Upper Jurassic Morrison Formation (Luo and Wible, 2005), has a skull length of ~ 1.2 cm, suggesting that Jurassic fossorial mammals may all have been of very small size, too small to have excavated the burrows in the Navajo Sandstone. If tritylodonts did excavate the burrows in the Navajo Sandstone, they possibly filled an ecological niche similar to that of modern dune-dwelling rodents or other mid-sized fossorial mammals, such as badgers.

Paleoecological and Paleobiological Implications

Identification of intensively-burrowed horizons in the lower part of the Navajo Sandstone near Moab, and inference of the likely burrow makers as tritylodontid cynodonts, are of both paleoecological and paleobiological significance. Loope and Rowe (2003) and Loope et al. (2004) identified thick, heavily-bioturbated intervals of invertebrate burrows and tetrapod footprints in the Navajo Sandstone over an area of up to 115 km^2 along the Utah-Arizona border. They inferred that such extensive bioturbation indicated relatively high rainfall necessary to sustain such high levels of biological activity (and inferred biomass), and thus posited two pluvial episodes during Navajo deposition that corresponded to the two heavily-bioturbated intervals. The burrowed horizons reported herein are just as heavily bioturbated and areally extensive as the bioturbated intervals reported by Loope and colleagues. The two intervals near the Utah-Arizona border, however, are 200-300 m above the base of the Navajo Sandstone, whereas the burrowed horizons near Moab are in the lower 26 m of the Navajo section. Although we doubt that the base of the Navajo Sandstone corresponds to a timeline (datum), it seems unlikely that the burrowed horizons near Moab correlate to the bioturbated horizons along the Utah-Arizona border. This suggests that episodes of heavy rainfall (pluvials) took place during Navajo deposition in addition to the two inferred by Loope and colleagues. With the added consideration of vertebrate trace fossils, the tritylodontid burrows indicate an even greater biomass in the Navajo depositional system than has been inferred from invertebrate trace fossils and tetrapod footprints alone. Indeed, an abundance of herbivorous tritylodontids implies sufficient vegetation to sustain these Early Jurassic ecological vicars of the desert-mammal niche.

Tetrapod burrows, especially those of rodents, are locally abundant in modern dune fields, especially in wet interdunal areas (Ahlbrandt et al., 1978; Gee et al., 2003). Such burrows indicate the presence of moisture and vegetation sufficient to sustain populations of desert-living rodents. The herbivorous, fossorial tritylodontids were very rodent-like in overall anatomy, so it is not surprising that at least some of them occupied niches now filled by desert rodents. Indeed, the Navajo burrow networks and mazes imply a degree of sociality in tritylodontids seen today, for example, in *Nannospalax* and other fossorial rodents.

Female *Nannospalax* construct breeding mounds above the ground surface during the rainy autumn and winter seasons, in order to situate nests and storage chambers above flood level. Thus, the mound is an adaptive response to climate. The mounds are "solid domes of earth containing a labyrinth of permanent galleries with hard, smooth walls" with a nest and several storage chambers in the midst of the galleries (Nowak, 1999). Mounds are typically 135-250 cm in diameter, rising 40-100 cm above the ground, but are much larger in areas prone to flooding (Nevo, 1961; Nowak, 1999). The large breeding mounds are usually surrounded by smaller mounds that the male mole rats construct, and connecting tunnels lead to and from the labyrinth inside the mound. These mounds are thus used for social interactions during the breeding season in a species that normally is solitary (Nevo, 1961; Nowak, 1999).

Old World moles (*Talpa* sp.) also build surface mounds about one meter high, containing the nest, in low-lying areas with high water levels (Nowak, 1999). If the structures in the Navajo Sandstone do represent burrow mazes within elevated mounds, they may suggest burrow system construction above ground level in the wet interdune environment to avoid waterlogging of burrows during the periods of heavy rainfall inferred for the Navajo paleoenvironment. Further work on these fossil burrows may help ascertain whether sociality or other behaviors, such as breeding, could be inferred for their excavator(s). Given that advanced cynodonts such as tritylodontids were very mammal-like in their skeletal anatomy, arguably endothermic, and may have (in some cases) approached a mammalian level of encephalization (Kemp, 1982), it is not surprising that their inferred behavior and social structure may have been very similar to that of some extant rodents.

This initial documentation of the Navajo Sandstone burrow casts near Moab is only the first step in the study of a very extensive and complex phenomenon. Many more burrows are known from the Navajo Sandstone than are discussed here, and similar burrows and burrow complexes are present in other Jurassic units on the Colorado Plateau, including the Middle Jurassic Entrada Sandstone and the Upper Jurassic Morrison Formation. Studies of these trace fossils will provide more insight into the behavior and distribution of Jurassic tetrapods than was previously known based only on their body fossil and footprint records.

REFERENCES

Ahlbrandt, T. S., Andrews, S. and Gwynne, D. T., 1978, Bioturbation in eolian deposits: Journal of Sedimentary Petrology, v. 48, p. 839-848.

Bromley, R. G., Curran, H. A., Frey, R. W., Gutschick, R. G. and Suttner, L. J., 1975, Problems in interpreting unusually large burrows; *in* Frey, R. W., ed., The study of trace fossils: New York, Springer-Verlag, p. 351-376.

Coombs, M. C., 1983, Large mammalian clawed herbivores: A comparative study: Transactions of the American Philosophical Society, v. 73, p. 1-96.

Eisenberg, L., 2005, Life in a Utah desert pond, 1850 million years b.p.: Canyon Legacy, v. 54, p. 21-27.

Ekdale, A. A., Bromley, A. G. and Pemberton, S. G., 1984, Ichnology: The use of trace fossils in sedimentology and stratigraphy. Tulsa, SEPM Sort Course 15, 317 p.

Gambaryan, P. P., 1960, The adaptive features of the locomotory organs in fossorial mammals. Yerevan, Izdatelstvo Akademii Nauk Armyanskoy SSR, 195 p. [in Russian]

Gee, C. T., Sander, P. M., and Petzelberger, B. E. M., 2003, A Miocene rodent nut cache in coastal dunes of the lower Rhine Embayment, Germany: Paleontology, v. 46, , p. 1133-1149.

Gobetz, K. E., In press, Possible burrows of mylagaulids (Rodentia: Aplodontoidea: Mylagaulidae) from the late Miocene (Barstovian) Pawnee Creek Formation, northeastern Colorado: Palaeogeography, Palaeoclimatology, Palaeoecology.

Groenewald, G. H., 1991, Burrow casts from the *Lystrosaurus-Procolophon* assemblage zone, Karoo sequence, South Africa: Koedoe, v. 34, p. 13-22.

Groenewald, G. H., Welman, J. and MacEachern, J. A., 2001, Vertebrate burrow complexes from the Early Triassic *Cynognathus* zone (Driekoppen Formation, Beaufort Group) of the Karoo basin, South Africa: Palaios, v. 16, p. 148-160.

Hasiotis, S. T., and Mitchell, C. E., 1993, A comparison of crayfish burrow morphologies: Triassic and Holocene fossil, paleo- and neo-ichnological evidence, and the identification of their burrowing signatures: Ichnos, v. 2, p. 291-314.

Hasiotis, S. T., 2002, Continental trace fossils. Denver, SEPM, 128 p.

Hasiotis, S. T., Wellner, R. W., Martin, A. J. and Demko, T. M., 2004, Vertebrate burrows from Triassic and Jurassic continental deposits of North America and Antarctica: Their paleoenvironmental and paleoecological significance: Ichnos, v. 11, p. 103-124.

Hembree, D. I., Martin, L. D., and Hasiotis, S. T., 2004, Amphibian burrows and ephemeral ponds of the Lower Permian Speiser Shale, Kansas: Evidence for seasonality in the midcontinent: Palaeogeography, Palaeoclimatology, Palaeoecology, v. 203, p. 127-152.

Hembree, D. I., Hasiotis, S. T., and Martin, L. D., 2005, Torridorefugium eskridgensis (new ichnogenus and ichnospecies): Amphibian aestivation burrows from the Lower Permian Speiser Shale of Kansas: Journal of Paleontology, v. 79, p. 583-593.

Hickman, G. C., 1990, Adaptiveness of tunnel system features in subterranean mammal burrows; in Nevo, E., and Reig, O. A., eds., Evolution of subterranean mammals at the organismal and molecular levels: New York, Wiley-Liss, p. 185-210.

Hildebrand, M., 1974, Analysis of vertebrate structure. New York, John Wiley and Sons, 710 p.

Hildebrand, M., 1985, Digging of quadrupeds; *in* Hildebrand, M., Bramble, D. M., Liem, K. F. and Wake, D. B., eds., Functional vertebrate morphology: Cambridge, Harvard university Press, p. 89-109.

Hotton, N. III., 1991, The nature and diversity of synapsids: Prologue to the origin of mammals; in Schultze, H. P., and Trueb, L., eds., Origins of the higher groups of tetrapods: Controversy and consensus: Comstock Publishing Associates, Ithaca, p. 598-634.

Kemp, T. S., 1982, Mammal-like reptiles and the origin of mammals. New York, Academic Press, 363 p.

Kielan-Jaworowska, Z., Cifelli, R. L. and Luo,. Z., 2004, Mammals from the age of dinosaurs. New York, Columbia University Press, 630 p.

Klappa, C. F., 1980, Rhizoliths in terrestrial carbonates: Classification, recognition, genesis and significance: Sedimentology, v. 27, p. 613-629.

Kocureck, G. and Dott, R. H., Jr., 1983, Jurassic paleogeography and paleoclimate of the central and southern Rocky Mountain region; in Reynolds, M. W. and Dolly, E. D., eds., Mesozoic paleogeography of west-central United States: Denver, RMS-SEPM, p. 101-116.

Kühne, W. G., 1956, The Liassic theapsid *Oligokyphus*: London, British Museum of Natural History, 149 p.

Lockley, M. G., 2005, Enigmatic dune walkers from the abyss: Some thoughts on water and track preservation in ancient and modern deserts: Canyon Legacy, v. 54, p. 43-51.

Lockley, M. and Hunt, A. P., 1995, Dinosaur tracks and other fossil footprints of the western United States: New York, Columbia University Press, 338 p.

Loope, D. B. and Rowe, C. M., 2003, Long-lived pluvial episodes during deposition of the Navajo

Sandstone: The Journal of Geology, v. 111, p. 223-232.

Loope, D., Eisenberg, L. and Waiss, E., 2004, Navajo sand sea of near-equatorial Pangea: tropical westerlies, slumps, and giant stromatolites; *in* Nelson, E. P. and Erslev, E. A., eds., Field trips in the southern Rocky Mountains, USA: Geological Society of America, Field Guide 5, p. 1-13.

Luo, Z. and Wible, J. R., 2005, A Late Jurassic digging mammal and early mammalian diversification: Science, v. 308, p. 103-107.

Luo, Z. and Wu, X., 1994, The small tetrapods of the lower Lufeng Formation, Yunnan, China; *in* Fraser, N. C. and Sues, H-D., eds., In the shadow of the dinosaurs: Cambridge, Cambridge University Press, p. 251-270.

Martin, L. D., and Bennett, D. K., 1977, The burrows of the Miocene beaver *Palaeocastor*, western Nebraska, U. S. A.: *Palaeogeography, Palaeoclimatology, Palaeoecology*, v. 22, p. 173-193.

McAllister, J. A., 1991, The lungfish Gnathorhiza and its burrows from the Permian of Kansas: Unpublished Ph.D. dissertation: Lawrence, University of Kansas, 170 p.

Morgan, G. S., Lucas, S. G., 2000. Pliocene and Pleistocene vertebrate faunas from Albuquerque Basin, New Mexico. In: Lucas, S. G. (Ed.), New Mexico's Fossil Record 2. New Mexico Museum of Natural History and Science Bulletin 16, pp. 217–240.

Nevo, E., 1961, Observations of Israeli populations of the mole-rat *Spalax ehrenbergi* Nehring 1898: Mammalia, v. 25, p. 127-144.

Nevo, E., 1999, Mosaic evolution of subterranean mammals: regression, progression and global convergence: Oxford, Oxford University Press, 512 p.

Nowak, R. M., 1999, Walker's Mammals of the World: Baltimore, Johns Hopkins University Press, 1936 p.

Odier, G. P., 2004, The Jurassic the rise of mammals after current discoveries in the Early and Middle Jurassic of southern Utah. Moab, Privately printed, 100 p.

Odier, G. P., Lucas, S. G., McCormick, T., Egan, C., 2004. Therapsid burrows in the lower Jurassic Navajo Sandstone, Southeastern Utah. Geological Society of America Abstracts With Programs 36, 67.

Ostapuk, P., 2005, On the trail of discovery: Canyon Legacy, v. 54, p. 28-30.

Rainforth, E. C. and Lockley, M. G., 1996, Tracking life in a Lower Jurassic desert: Vertebrate tracks and other traces from the Navajo Sandstone: Museum of Northern Arizona Bulletin 60, p. 285-289.

Smith, R. M. H., 1987, Helical burrow casts of therapsid origin from the Beaufort Group (Permian) of South Africa: Palaeogeography, Palaeoclimatology, Palaeoecology, v. 60, p. 155-170.

Stein, B. R., 2000, Morphology of subterranean rodents; *in* Lacey, E. A., Patton, J. L., and Cameron, G. N., eds., Life Underground: The Biology of Subterranean Rodents: Chicago, University of Chicago Press, p. 19–61.

Sues, H-D., 1986, The skull and dentition of two tritylodontid therapsids from the Lower Jurassic of western North America: Bulletin of the Museum of Comparative Zoology, v. 151, p. 217-268.

Voorhies, M. R., 1975, Vertebrate burrows; *in* Frey, R. W., ed., The study of trace fossils: New York, Springer-Verlag, p. 325-350.

Winkler, D. A., Jacobs, L. L., Congleton, J. D. and Downs, W. R., 1991, Life in a sand sea: Biota from Jurassic interdunes: Geology, v. 19, p. 889-892.

Young, C. C., 1947, Mammal-like reptiles from Lufeng, Yunnan, China: Proceedings of the Zoological Society of London, v. 117, p. 537-597.

FIGURE 1. Index map of Utah and measured stratigraphic sections showing the two burrow cast localities in the Navajo Sandstone documented here.

FIGURE 2. Selected photographs of NMMNH locality 5680, tetrapod burrow casts in the Navajo Sandstone. A, Overview of the locality showing main burrowed horizon (b). B-D, Mounds (m) formed by burrow mazes. E, Differentially weathered network of burrow casts. F, Complex chamber with multiple entrances (e).

FIGURE 3. Selected photographs of NMMNH locality 5681, tetrapod burrow casts in the Navajo Sandstone. A, Overview of the locality showing main burrowed horizon (b). B-D, Differentially weathered burrow casts; note multiple entrances and cul-de-sacs, especially evident in C. E, Rhizoliths (r) and burrow casts (b) together in cross-sectional view of outcrop; note differential preservation of rhizoliths. F, Typical, downward-branching rhizolith.

FIGURE 4. Selected tetrapod burrow casts and a rhizolith from NMMNH locality 5681. A-B, NMMNH P-45278. C, NMMNH P-45279. D-E, NMMNH P-45280. F-G, NMMNH P-45281. H-I, NMMNH P-45277, rhizolith fragment in lateral (H) and cross-sectional (I) views.

FIGURE 5. Lateral view of the skeleton of the tritylodontid cynodont *Oligokyphus* (from Kühne, 1956).

TABLE 1. Characteristic features of vertebrate burrows.

1. distinctive architectural morphology (e.g., entrance shaft leading to primary tunnel with secondary and tertiary branches)
2. subcircular or elliptical burrow cross sections of consistent diameter
3. cut across bedding
4. burrow fill contrasts with host strata (this may only be detectable at a microscopic level, cf. Gee et al., 2003)
5. preferentially preserved
6. possess linings
7. distinctive surficial morphology, including longitudinal ridges and/or paired grooves that may be interpreted as scratch marks, beak marks, or tooth marks
8. one or more entrance holes to surface
9. variable branching pattern
10. multiple terminal chambers, each having multiple entrances
11. shallow vertical or low-angled shafts lead from ground surface to low angle ramps, helical tunnels, or an underground maze

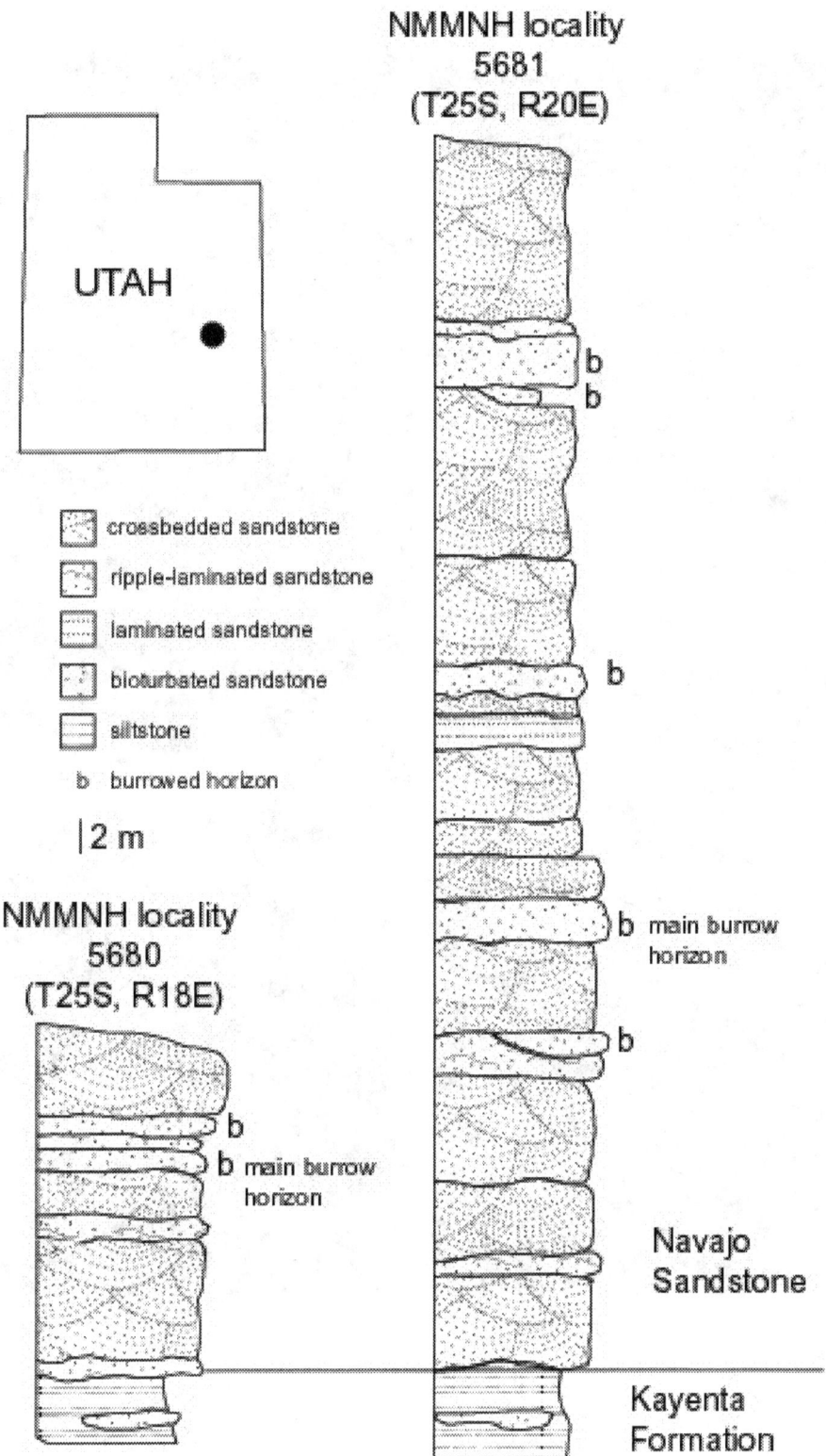

UTAH

NMMNH locality
5681
(T25S, R20E)

crossbedded sandstone

ripple-laminated sandstone

laminated sandstone

bioturbated sandstone

siltstone

b   burrowed horizon

|2 m

NMMNH locality
5680
(T25S, R18E)

b

b   main burrow
horizon

b

b   main burrow
horizon

b

b

b   main burrow
horizon

b

Navajo
Sandstone

Kayenta
Formation

Figure 1

Figure 2

Figure 3

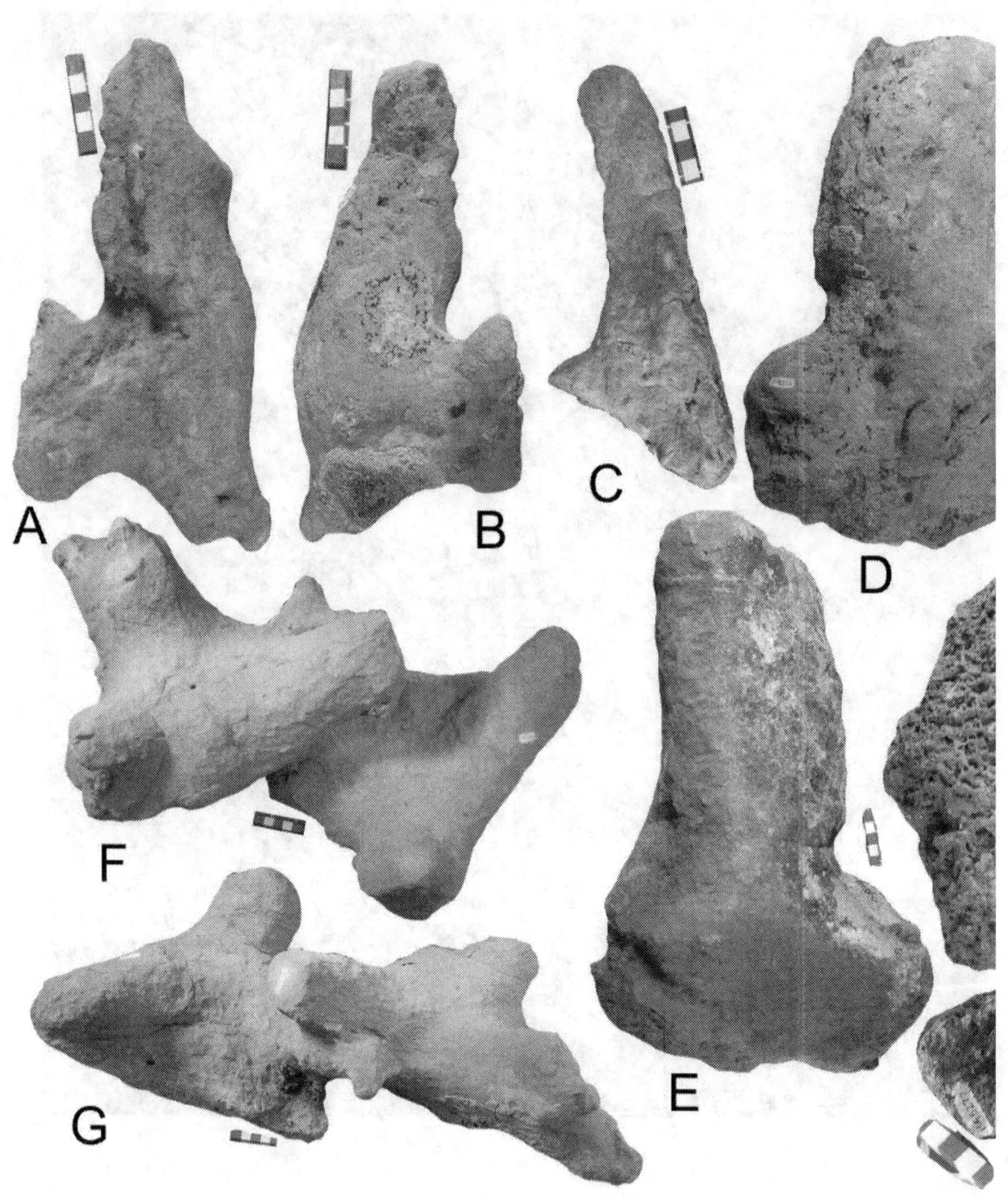

Figure 4

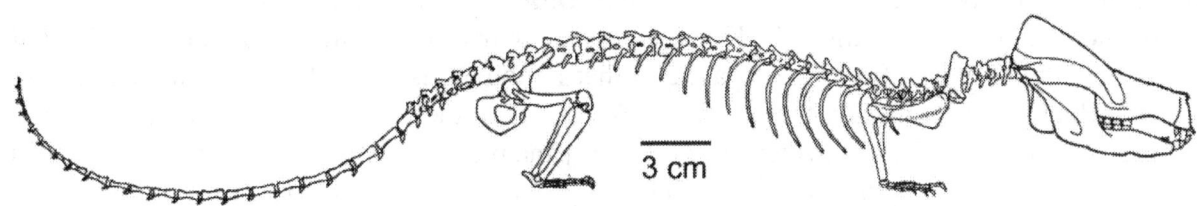

3 cm

Figure 5

Author's note:

The above manuscript is the extension of Abstract # 77429 presented at the November 2004 GSA Annual Meeting. It was completed May 20, 2006, and successfully peer-reviewed by July 15, 2006. It's scheduled to be published in Bulletin #30 of the New Mexico Museum of Natural History & Science under a study and review of the Triassic-Jurassic Boundary. This manuscript represents the first scientific evidence supporting these Navajo burrows as mammal-made. The first step in the long process to document and vindicate the discovery discussed at length in this book.

This burrow, today nicknamed the 'Curecanti Burrow', was uncovered in 2005 by NPS paleontological technician Alison L. Koch, and NPS archeologist Forest Frost in the Salt Wash Member of the Morrison Fm (Late Jurassic in that region). It is located in the federal Curecanti National Recreational Area adjacent to the Black Canyon of the Gunnison National Park, Western Colorado. Ascribed by the discoverers as 'tentatively mammal-made', it was submitted to Stephen T. Hasiotis, Assoc. Professor of Geology, University of Kansas, who documented it as vertebrate-made, reptile or mammal. Based on this information, and her own, Alison Koch wrote a paper, co-authored by Forest Frost that includes Kelli Trujillo, Laramie, Wyoming, as assistant to the discovery. This paper came to the attention of Spencer Lucas, Curator of Geology and Paleontology at the New Mexico Museum of Natural History & Science, Albuquerque, New Mexico, in late May 2006. He transmitted this information to me in order to check the location and veracity of this burrow with the Park's scientific staff. On June 6, 2006, Forest Frost forwarded me five pictures he took of the site that very morning, before fast rising water inundated the burrow for the Summer season. Based on these pictures, I determined the site to be a badly eroded and battered Class III burrow similar to many in the Navajo Sandstone (Early Jurassic). These five pictures, courtesy of Forest Frost, are included for visual identification.

The 'Curecanti' burrow is not the first one uncovered in the Late Jurassic of North America. The first one is, or appears to be, burrow-casts uncovered in 2002 by Anthony Martin, University of Georgia, in-around Egg Mountain, near Bozeman, Montana. However, Anthony Martin's documentation is insufficient to confirm these burrow-casts as mammal-made, although they are or appear to be, based on the Navajo burrows today unconditionally of mammalian origin. The second one was uncovered in the Henry Mountains (100 miles west of Moab, Utah, as the crow fly) and documented in 2004 by Stephen Hasiotis, Robert W. Wellner, Anthony Martin, and Timothy M. Demko in an Ichnos paper titled: 'Vertebrate Burrows from Triassic and Jurassic Continental Deposits of North America and Antarctica: Their Paleoenvironmental and Paleoecological Significance'. Based on personal correspondence between Stephen Hasiotis and I, the Henry Mountain and Curecanti burrows have similar architectural morphology, are vertebrate-made, and again based on the Navajo burrows, are most likely of mammalian origin. However, for the record, Spencer Lucas has challenged (May-June 2006) the Henry Mountain burrow as vertebrate-made, instead suggesting erosonial features, concretions, etc., of geologic origin. Hereby the importance of the 'Curecanti' burrow. Most likely this burrow is going to play a pivotal role in the 'deciphering' of Late Jurassic burrows. These burrow-casts were made in sedimentary deposits thus differ from the ones documented in aeolian deposits (Wingate, Navajo, Entrada, etc., in southern Utah). This differentiation, small to large, is still in its early stage as per this date. As soon as a morphological and preservation pattern is established for these Late Jurassic burrows we can expect to find many more in North America and elsewhere.

"Curecanti Burrow"

photo: Forest Frost

photo: Forest Frost

photo: Forest Frost

photo: Forest Frost

photo: Forest Frost

Abstract

Preliminary report on dewatering pipes in the lower part of the Lower Jurassic Navajo Sandstone, Canyonlands National Park, southeastern Utah: implications for pluvial episodes and the occurrence of lakes, trees, and mammal burrows

Odier, Georges[1], **Hasiotis**, Stephen[2]*, Rasmussen, Donald[3], McCormick, Tamsin[4]
[1]115 W. Kane Creek Blvd. # 29, Moab, UT 84532, godier@preciscom.net; [2]University of Kansas, Department of Geology, 1475 Jayhawk Blvd., 120 Lindley Hall, Lawrence, KS 66045-7613, hasiotis@ku.edu; [3]Plateau Exploration Inc. 1450 Kay St. Longmont, CO 80501-2427; [4]Plateau Restoration, P.O. Box 1363, Moab, UT 84532; *Presenter

Numerous dewatering pipes crosscut eolian dune-interdune deposits in the lower part of the Lower Jurassic Navajo Sandstone in and around Canyonlands National Park, near Moab, Utah. The Navajo Sandstone was deposited as a vast erg system over northeastern Arizona, southwestern Colorado, and most of Utah, and unconformably overlies the Upper Triassic Chinle Formation.

The dewatering pipes range in size and distribution greatly, and are so abundant that they form fields of dewatering pipes. The smallest pipes are 4-200 mm in diameter. The largest pipes are 1-4+ m in diameter. In some cases pipes are so large and oddly shaped that the exact size is indeterminate. Smaller diameter pipes originate from larger diameter pipes. The internal morphology includes concentric sediment layers and rotated blocks of cohesive laminated sediments.

The fields of de-watering pipes appear to terminate at several levels associated with vast interdune deposits traceable for several kilometers. Interdune deposits contain abundant evidence of pedogenic modification and wet ecosystems. Pedogenesis is indicated by massive fine-grained quartz sandstone and pale reddish brown siltstone and mudstone with 3- to 100-mm-wide rhizoliths, bleached rhizohalos, calcareous rhizocretions, bored steinkerns and permineralized tree trunks and roots, 3- to 5-mm-wide adhesive meniscate burrows, 5-mm-wide subvertical burrows, and large-diameter subhorizontal burrows with T- and Y-shaped intersections and switchback tunnels that form pseudo-spiraled ramps. The traces represent activity of soil bugs, beetles, social therapsids or mammals, and vegetation. The association of pipes, interdune deposits, and pedogenesis indicate that local-regional groundwater recharge supported this ecosystem during pluvial intervals.

Author's notes:
The above Abstract, completed on June 26, 2006, will be submitted to the Topical Session 48 (T48): Geology in National Parks: Research, Mapping, and Resources Management, at the 2006 Annual Meeting of the Geological Society of America (GSA). Stephen Hasiotis Phd, will be the Presenter.

www.ingramcontent.com/pod-product-compliance
Lightning Source LLC
Chambersburg PA
CBHW081119170526

45165CB00008B/2490